Science and Spirituality

Cultivating

THE SPIRIT OF
UNITY-AND-DIVERSITY

by

Mary A. Mann
Leland P. Stewart
and compilers

authorHOUSE™

1663 LIBERTY DRIVE, SUITE 200
BLOOMINGTON, INDIANA 47403
(800) 839-8640
WWW.AUTHORHOUSE.COM

First published by AuthorHouse 11/17/04

ISBN: 1-4184-9295-7 (e)
ISBN: 1-4184-9294-9 (sc)

Library of Congress Control Number: 2004096542

Printed in the United States of America
Bloomington, Indiana

This book is printed on acid-free paper.

for your copy go to
www.scienceandspirituality.com.

Including an Interfaith Celebration Guide

BOOK TWO   2004

Cover:
The spiritual logo of Unity-and-Diversity World Council
Poem: Percy Bysshe Shelley

Compiled by:

Monique Baudaux Andersen
Rene Crawford
Rev. Shirley Leighton
Mary Anneeta Mann, Ph.D.
Rev. Elizabeth Stewart
Rev. Leland P. Stewart, B.S.E., S.T.B.

BOOK ONE was compiled in 1994 under the title of

A UNITY-AND-DIVERSITY SPIRITUAL CELEBRATION

UNDERSTANDING "UNITY-AND-DIVERSITY"

The term "unity-and-diversity"
is intentionally hyphenated in order to show that it is one concept
rather than separated, unconnected ideas; namely "unity" and "diversity."

Democracy is based upon a pluralism of races, cultures and religions
being contained within one society.

It is a very rare and fragile concept,
and one which needs to be cherished and preserved.

Even in an officially democratic society, such as the United States of
America,
it is often forgotten or misused.

In recent times, it has increasingly become the property of the world as a
whole,
yet it is far from being understood and appreciated.

One of the purposes of this book therefore, is to help people
understand
"unity-and-diversity"
and how it applies to an increasingly pluralistic world.

Everyone who is

seriously involved

in the pursuit of science

becomes convinced

that a spirit is manifest

in the laws of the

universe

Albert Einstein

TABLE OF CONTENTS

SCIENCE AND SPIRITUALITY

Spirit is One, Paths are Many

 Baha'i

 Bawa Muhaiyaddeen

 Brahma Kumaris

 Buddhism

 Cao Dai

 Chinese Taoism Confucianism

 Christianity

 Christian Science

 Hinduism

 Islam

 Jainism

 Judaism

 Native American

 Religious Science

 Scientology

 Shintoism

 Sikhism

 Theosophy

 Zoroastrianism

 All Others

WE

ARE

THE CHILDREN

OF THE WORLD

WE ARE ONE

WE ARE

ONE

DEDICATED TO

People sharing spiritual journeys

celebrating the same unity

participating in diversity

developing a global ethic

SCIENCE AND SPIRITUALITY
THE PURPOSE OF BOOK TWO

Book One in this series was called *Unity-and-Diversity Spiritual Celebration Guide*. Describing a Unity-&-Diversity Spiritual Celebration, its purpose was also to provide materials for Interfaith Celebrations through the use of a *Candlelighting Ceremony*, with messages from the different faiths on a pre-selected theme, *Meditation,* and *Responsive Readings*.

Science and Spirituality is designed for a much more general purpose. It is intended to encourage people to meditate upon their daily lives and to look at the course of our present civilization and their participation in it. It includes a review of the world's great religions and modern spiritual movements, as well as brief articles on the spirituality, cultures and traditions of Africa, North America and Australia. It includes a *Declaration of Interdependence*; a *Global Ethic*, 1993 Parliament of the World's Religions; *Universal Declaration of Human Rights*, 1948 General Assembly of the United Nations; *The Earth Charter*, 2000 Peace Palace, The Hague; *Civilization Timeline*, from 3000 B.C.E. to the present; readings for *Candlelighting and Meditation, Responsive Readings* and *Songs of Inspiration. The Interfaith Celebration Guide* mirrors the earlier spiritual guide but draws upon greatly expanded readings from diverse sources including the world's religions and literature, as well as quotations from great philosophers, scholars, writers, scientists and mystics of our civilization from ancient times to the present day. One of the major goals of this effort is to help us all to understand the intrinsically harmonious relationship among the world's faiths and the great thinkers and seekers of all time. We believe that this effort will help lead us to the moral imperative to act upon this understanding and come together in creating a global civilization.

Peoples' activities are shaped by the age in which they live. The twentieth century was largely dominated by the arrogance of human beings in power. People killed each other in record numbers. Boundaries of nations were changed and changed again. New nations were created, and there was such a dispersion of human migration that there are few nations left with ethnically homogeneous populations. The damage to the environment was greater in one century than in all of the centuries preceding it.

On the positive side, science and technology advanced greatly, until mortals almost comprehended the workings of the universe itself. The beliefs of the historic religions have been tested beyond their breaking point but have survived. Many new spiritual groups and religions have begun.

Now mortals are beginning to understand their true role in the workings of a universe where spirituality, religion and science coalesce and where philosophy returns to its own roots in describing this phenomenon or this paradigm. We all may now gaze in awe at the magnificence of the cosmos being revealed to us through science. The magnitude of our potential role could serve to focus our minds and hearts on actions which represent movement away from destruction and toward our life epiphany, individually, collectively and communally with the earth and the environment.

Book Two, *Science and Spirituality***,** is a response to this awareness at the dawn of the twenty-first century, an age which requires pioneers of human thought and human faith.

As the impulse to understand this coalescence generates more human activity, **Book Three** will follow. The dedication to understanding the workings of the universe needs the resources of every human culture living and dead as well as every effort that our generation can muster. If the emerging civilization is to be the one to survive, it will have to move humanity beyond the polarizations and destructive impetus of the early years of the twenty-first century. It will have to recognize this impetus as threatening the entire emanating civilization. It will have to comprehend this on the pulses, and then consciously move the course of this global civilization into harmony with the workings of the universe. When this happens, creativity and scientific advances may be mined for the magnificence of their revelations concerning cosmic harmony.

As cosmic harmony is accepted as the way of our world, poverty is addressed and eliminated. There is food enough on this planet to feed everyone. Human dignity is held inviolate. Children learn at their mother's knee that the greatest human virtue is *reverence for life*, reverence and respect for their own life, their own body and their own spirit. As they secure this for themselves they understand that empathy for others is inherent in it. Hatred becomes impossible. "Enemy" becomes an emotive utterance with no real meaning. Conflicts are not deadly. There is more challenge and excitement in resolving differences peacefully. Creativity becomes an essential part of daily life. Governments exhibit gender equality. Artists are invited to participate in government, where they are recognized as the "legislators of the world", no longer unacknowledged. The global family includes all animate and inanimate beings as it does in the ancient cultures. The refinements of "civilization" are: awareness of spiritual belongingness and responsibility, freedom from poverty, the benefits of safety, security, healthful life styles based upon ecological integrity, freedom of speech and freedom to worship, freedom to participate in social activities and world affairs, freedom to have differences and freedom to resolve them without bloodshed.

The ultimate goal of all of the books in this series of *Science and Spirituality* is to hold up a vision of cosmic harmony until it becomes the reality that all of the great faiths and the great minds of the world have deemed it to be.

COMPILERS' NOTES

We all feel privileged to have worked on this book. It has been a wonderful opportunity to visit with the great religions of the world as well as great minds from science, literature and forward-thinking people who are our contemporaries.

We wish to thank all of those members of our organization who have searched through their records and sent us quotations from their favorite authors or religious leaders. These include among others: Deborah Bloedorn, Swami Brahmavidyananda, Doris W. Davis, Al Duffy, Keki B. Gandhi, Carl Hult, Rabbi Moshe Halfon, Richard Kuznetsky, Sam Sohrab, Rev. Jill Soyars and Rev. Alfred Tsuyuki. Special gratitude goes also to Dwight Stone, the music coordinator, Janice Kalec for books on Community and Native Americans, Norma Bigtree Groverland and Gregory Schaff, Ph.D. for 'The Great Law of Peace and the Constitution of the United States of America", and Penny Colburn and Rev. Jill Soyars for editing assistance.

The first book began using an oral tradition of quotations from the great religions, which were used in the Candlelighting Ceremony. Later, quotations were found from other sources. The translators of the books of the New and the Old Testament in the King James Version of the English Bible were also poets in their own right and raised much of the English Bible to the level of great literature, so from this perspective the names of books may be used individually thus expanding the reach of Christianity.

On some occasions, gender equality has been imposed on the original quotation. Sometimes the text has been altered to update older usages of words. Some changes may be more noticeable in poetry quotations. Even Shakespeare has had some of his language altered to bring it into the twenty-first century. With these changes, we have been guided by the great bard himself, *my words shall live despite thy scythe and thee*, in that we felt that these pearls of wisdom deserve entry into our present world for their *essence*.

We also wish to ask for the reader's indulgence concerning our deficiencies in knowledge and perspective. We have all done the best we could with the resources available to us. There are gaps in knowledge, understanding and cultural awarenesses, which we recognize. But it is the yearning to see this civilization better equip itself with the tools of survival that has urged us forward over the last ten years. It is our hope that this kind of work will be carried on by others.

Our appreciation is given to all authors and publishers who allowed us to reprint their material in this book. Should there be any quotation or extract included whose work appears without proper arrangement or acknowledgment, please advise us and it will be corrected in the next printing. We do not desire to use anyone's material without permission or proper credit. We would also ask for your indulgence where we may have erred or not listed an author in the index. If you have material that you feel may be valuable to us for the next edition, you may send it to Dr. Mary A. Mann, P.O. Box 9356, North Hollywood, California, U.S.A. 91609. We will accept it with gratitude.

THE SPIRIT OF UNITY-AND-DIVERSITY

Igniting the Flame that Gives Life to the New Person and Civilization

The emergence of a unity-and-diversity movement that fires the flame of faith in the future of life on planet earth and beyond, is what is now required to transcend the current stalemate, which is symbolic of our arrival at a global crossroads.

The key phrase is "unity-and-diversity among all peoples and all life". It is the coming into visibility of the merging of science and religion, the proclaiming of a new basis of life which shall be for all peoples. It will unite the left and right hemispheres of the brain as we discover our wholeness. It will enlarge the circle of that wholeness to include all races, all religions, all nations and cultures of this planet. It will prepare us for moving out into space in a creative and responsible way.

"Unity-and-diversity" is key to the new world because it is necessary to respect the diversities of race, religion, culture, and understanding before the unity will come. On U.S. coins this is called *e pluribus unum*. It means *one for all and all for one*. It is the mystical principle of *the one and the many*. Required for the new civilization is the capacity to listen to differences as well as to proclaim the new understanding, to have *reverence for life* and to experience the reality of the many-faceted diamond which is life itself.

The role of such a unity-and-diversity movement is to provide the vision and the dynamic equilibrium needed for launching the new civilization. It is the basis for the moment-to-moment aliveness, the transformational energy required to move through crisis and disillusionment into the vastness of the universe that lies before us. That universe is both within us and around us; with all of our exploration to date, we have just begun.

In the light of the new universal consciousness, each moment is a joy to behold because it offers an opportunity for growth, for seeing life possibilities anew.

Welcome to the moment of truth in our life, when we all let go of our limited views and accept our universality. Welcome to our wholeness, our cosmic identity.

WELCOME HOME!

Rev. Leland P. Stewart, B.S.E., S.T.B.
Founding Visionary, Unity-and-Diversity World Council

SCIENCE AND SPIRITUALITY - a philosophical discourse

Science operates in the field of knowledge, which is human, *spirituality* in the field of understanding, which is divine. These two fields overlap, as every human being partakes of the divine.

The field of knowledge consists of facts, things done or performed, particular truths known by actual observation or authentic testimony. Truth for science is the coincidence of a subject with its object. What can be proven scientifically expands as far as the mind of the scientist can make an induction or an intuition and then back it up with fact. For instance, 'the world is flat' was proved to be an untruth when Columbus sailed around it. 'The world is round' then became the truth. A fact however, a datum of experience, is limited by the capacity of the human mind which is finite.

In spirituality on the other hand, scientific affirmation is possible but never scientific confirmation as its field of understanding partakes of the infinite. The infinite is the only thing which is non-exclusionary. We can never prove that something which is exclusionary is true for all time. The finite mind cannot comprehend the infinite. What the finite mind can do is partake of TRUTH, an inherent characteristic of the infinite once we can figure out how to tap into or access the field of understanding.

In *spirituality*, what we are seeking then is not a 'fact' but a principle of operation of the universe. We are seeking to understand the movement of the Life Force of the universe, and everything knowable about it which will keep expanding all the time, as well as what can be apprehended of the infinite.

This 'principle of operation' or Life Force has been called 'God' but the use of an apparent substantive has caused some religions to attempt to concretize this 'God'. There are two fallacies here. One fallacy is the apparent attempt to make the infinite finite and the second fallacy is the linguistic one which lures a moving principle of operation, or Life Force toward a concrete symbol.

*If the religions can come to this scientific awareness of the **Life Force** as the **moving God** then many apparent inconsistencies can become comparatively insignificant.*

Ideally each religion is the means whereby a human being can at least seek - through ritual and a certain measure of conformity - this field of understanding where TRUTH becomes real and the unity all-encompassing.

There can be many diverse journeys that people take on their way to this oneness. The journey itself is significant for each individual. The study of other religions gives knowledge. Comparisons also add to quality for humans individually as they draw comfort from the spirit of community, while the contemplation of one's own religion or one's own journey, accesses *spirituality.*

Let us look at what the fields of knowledge (*science*), and understanding (*spirituality*), have in common. The Life Force of the universe can possibly be termed the supreme general law common to both science and understanding. The outer limits of science are defined by the capacities of the human mind and so are being expanded all the time. Some truths become superceded by others as the seekers bring particular laws to general laws and general laws to the Life Force of the universe. The field of understanding that goes beyond science is nevertheless rooted in science in the sense that an apprehension of the principle of operation of the universe is a leap of faith from scientific truth. It requires spiritual activity on the part of the human being. It requires effort. This effort is the activity of each individual spirit to seek its source which is a common source and to imagine its destiny or its end which is a common destiny or end. Thus the only unique thing is the individual human being, for the source of the spirit is a common source to all humans and the destiny is a common destiny. The first hurdle for religions to overcome is that of the tendency to concretize a principle of operation or the Life Force and to consider exclusionary, the infinite which is non-exclusionary.

The next issue is that of the concept of *good*. In many ways 'good' is like a truth. A truth is the coincidence of a subject with its object. 'Good' is the close approximation of grader and graded. A good pen is one that writes well but what is a good person? What is the principle of grading for a good person? While we can and we must codify modes of conduct that will lead us toward continuing life on this planet, what we use to grade human conduct again is this same principle of the Life Force of the universe that we used to understand our participation in TRUTH.

9

In this framework of reference, 'evil' needs to become synonymous with 'bad' in the sense of not matching the mode of 'good' in this grading sense.

Religions have struggled to understand the concept of a 'God' that can encompass both so-called 'good' and 'evil', or 'good' and 'bad'. When we consider the mode of operation of the universe however, the source is a common source and the destiny is the ever-changing but ever fulfilled and thus ever whole, universe. There is this inevitable drive toward fulfilment and thus perfection of the whole. The 'grader' then or the 'standard' or the 'model' is not a concrete but a moving destiny of fulfilment. Each part is integral to the whole and each part moves toward its own self fulfilment while deferring to the inexorable drive of the destiny of the whole. This principle of operation of the universe, other wise described as the energy drive or Life Force is the most powerful of all energy drives in the universe in the sense that it prevails over all others. For instance if, under certain conditions, the earth's layers will shift, then an earthquake will occur in a certain place. Good and upright people may have built a house there. The ensuing destruction cannot truthfully be called 'bad things happening to good people'. Rather it is the prevailing of the principle of the Life Force of the universe despite our ignorance of it. It is the divine GOOD, and human beings partake of it as they accept it, recognize, or attempt to understand its power, and live their lives in harmony with it.

If we wish to try to describe or codify conduct of human beings, or human morals, first we have to try to intuit and/or understand how the Life Force of the universe might be working based upon the uttermost limits of what science can reveal to us. This shows that science and spirituality are kindred spirits in the seeking with science ending at the limits of knowledge and spirituality continuing on faith in the quest toward reaching the infinite and in doing so, participating in TRUTH.

Participating in TRUTH might be termed the reward or the success of life. As more people reach out toward this participation, so will our collective understanding grow and we will be in a stronger position to codify the new global covenants, or the new global morality, or the new moral directives for the actions of our lives. These will lead us away from the destructive course we embarked upon and toward the preservation of our species and the planet as they align themselves with the energy drive toward fulfilment and the on-going perfection of the whole. For example, the quality of beingness equals *spirituality* and the quality of action equals **ethics** or the *science* of morals.

At the moment of choice of action, each one of us chooses from an array of available options. If a mother sees a child in danger, she will run in front of traffic to save it. She will muster superhuman strength to lift a weight. Rescuers risk their lives in fires, floods and other disasters and act to save the life of another human being or another creature. This impulse is rooted in *spirituality*. The impulse is the will to live and the will to preserve life. The action itself falls in the area of ethics. An ethical person makes the most significant choice of saving. A less ethical person makes another choice from the available options. When someone is being attacked in a street, some people pass by without attempting to help. If we wish to ensure that human beings globally will choose at a gut level to save the life of another rather than turn away, we need to be sure firstly that the will to live and preserve life is an inherent characteristic of the state of being human and secondly that this natural but yet ideal impulse is allowed free expression and is recognized and accepted as *ethical*. For instance, if someone is being beaten on a street, those that go to help need to feel that they will not be punished for it by being shot or being the subject of retaliation. Governments themselves need to adhere to a common global ethicality which is based on *reverence for life*.

Specifically, if we are to foster life preservation, instruments and methods of domination and death need to be phased out while reinforcement of the will to live is offered through example, as well as opportunities for actions which manifest this ideal human behavior. For Aristotle the science of being altogether included the whole universe. The object of science is for mortals to acquire knowledge that enables them to become the conscience of the universe. On becoming the conscience of the universe and consciously participating in the divine they recognize the divine in every human being regardless of that human being's actions which may be ethical or unethical; and they treat that human being with the respect and dignity inherent to its being, or with the awareness of *reverence for life*. They accept responsibility of being the primary students and servants of the Life Force of the universe and as such, custodians of planet earth and her environment up to the uttermost and ever expanding limits of the finite mind and being.

Mary Anneeta Mann, Ph.D..

11

SEEKING GLOBAL HARMONY

Harmony is our birthright. Harmonious people have consciously grasped the divine thread of their existence and cherish it above everything else in their lives. Emanating from their own hearts, harmonious people have no enemies. Harmony is not static. It moves as our awarenesses move. Harmony requires courage to make personal adjustments, to keep moving toward understanding the motives of those who feel justified in hating and in having enemies. Harmonious people bear the responsibility to move beyond their own personal harmony and to actively participate in melting the hatred that fuels the concept of 'enemy', because it has been proved to be un-scientific and violating the spirit of the universe which gives to each and every being the inherent right to 'be'.

nine eleven and the perception of nine eleven

How we perceive anything shapes our thought and our thoughts shape our own action. As our own life journey merged on that fateful day with the life journey of the great country we call America, the horror of the spectacle challenged our perception of what really happened. What did the twin towers really represent? Did they represent the same thing to all people? We can all say with certainty that they did not.

Nine eleven may go down in history as the great test of our humanity and our civilization. Everyone in the world was affected by nine eleven in some way.

the choices offered after 911

the easy way...
911 made it easy to have enemies and easier still when the new enemy has no face, as 'terrorism". The faces of those who did the deed are mingled with the others. The faces of some of the planners are still among us.
... to make the assumption that 'freedom', 'way of life', 'what we stand for' was under attack and ...to have thoughts of fear and declare an enemy

was this a beginning?

the difficult way...   ...   ... and the way not taken...

The semblance of 'terrorism' challenges us all to survive it. How can we find out what was the true cause of it? Are the assumptions of the 'easy way' valid? There was a trail of terror before this and just as Karl Marx and Hitler laid out their assumptions about their contemporary societies, before they acted so did Osama Bin Laden

...to exercise due diligence in understanding what the assumptions were and how they might be clarified without bloodshed.

was this in the middle of a journey?

We understand the humanity in those reacting to 911. They are our own, most of them our native sons and daughters once again laying down their lives for their country's call, with the utmost faith. Greater love has no one than this but in the heart of every sacrifice there is an instilled 'enemy' generated by someone's hatred which, history has proved is not inherently in the heart of the citizen soldier. *Every such soldier and every soldier's family becomes victimized in war.*

The thought of 'enemy' and the action of war took us all back in time to the 20th century and not forward into the 21st. Great progress had been made in the 50 years since the end of *World War II* and 75 years since the end of *World War I*, to move our global thought beyond the necessity of war toward the necessity of a global community where those who cause destruction are judged and sanctioned by the ***will of the world and not the terror of war.***

Now, harmonious people through the tears and pain that are threatening their birthright are being compelled to seek the understanding that can lift us all out of a quagmire of hatred which is being fueled by the deception inherent in taking the easy way out of an immediate crisis that results in perpetual dis-harmony for all those caught up in the glue of it.

A pattern of hating has defined much of our civilization as shown in the struggles for dominance in the CIVILIZATION TIMELINE in this book that goes back to the building of the pyramids and forward to the 'war on terror.'

the divine way ...

Global harmony now requires that we pay more attention to the things that we believe, for what we believe motivates our actions. Will we allow our civilization to go to its destruction because we refused to look with clear eyes at what we believe ourselves and what we believe that others believe?

..to have faith in the creativity of the Life Force the moving God, to have the self-confidence to search our own hearts, our own beliefs, our own understanding of the workings of the universe and to move ourselves as the conscience of the planet into harmony with it knowing that its primary directive is reverence for life and to act accordingly.

how to touch the divine and how to act upon its directives

The great religions of Zorastrianism, Judaism, Christianity and Islam all talk about good and evil, but God is the judge. How often are we all tempted to be a self-appointed judge of good or evil? How often are we all tempted to use the words good and evil to refer to people instead of the actions of people and with what humanly divisive results.

These same great religions in their practices have all denied equal participation to women. This is unscientific, as the organic workings of the universe relentlessly require that each and all components in the organic whole participate to the full extent of their capabilities.

Most people in the world believe that they cherish democracy. In an election year naked democracy would require educational booklets distributed to every person of voting age describing what each party or candidate is offering as their version of government of the people and for the people. Do we have the courage to open debate about embellished versions of advanced democracies? those that require personality contests, misleading and often inaccurate sound-bites of persuasion, paid for by capitalists who by definition owe their first allegiance to the development of financial and industrial WEALTH? It is unscientific to market image and merchandise which has no component of value inherent in it. Can or should harmonious people mediate such debates in the interest of gaining the assurance that the people are truly electing those who express their truly educated collective will?

As we come to grips with these issues, we will all know that global harmony is indeed within our reach in the twenty first century, the century in which humankind is finally able to learn from past missteps occasioned by such a tragic loss of life of those we have placed on the front lines of our conflicts that to date have not secured our enlightenment nor brought us the peace for which all the wars have been fought.

As we look at the comparative beliefs that follow, we can clearly see how our kinspeople through the ages have accepted their religious dichotomies on unexamined 'faith' that is often not the truth as written by their founder.

Where there is truth to the dichotomy, we must now ask ourselves why we continue to accept the concept of 'enemy' when it may be our next-door neighbor who practices a different religion.

These Comparative Beliefs are taken from eleven major religions and a twelfth from other faiths. We have obtained this material from the Himalayan Academy Publications in Hawaii and from Brady's Peace Cards in Inglewood, California. We are using their information with thanks.

M.A.M.

COMPARATIVE RELIGIOUS BELIEFS

TAOIST BELIEFS

1.I believe that the Eternal may be understood as the Tao or "Way", which embraces the moral and physical order of the universe; the path of virtue which Heaven itself follows; and the Absolute-yet so great that "The Tao that can be described is not the Eternal Tao".

2. I believe in the unique greatness of the sage Lao Tsu and in his disciple Chuang-Tsu.

3. I believe in the scriptural insights and final authority of the Tao-Te-Ching and in the sacredness of Chuang-Tsu's writings.

4. I believe that we align ourselves with the eternal when we observe humility, simplicity, gentle yielding, serenity and effortless action.

5. I believe that the goal and the path of life are essentially the same, and that the Tao can be known only to exalted beings who realize it themselves - reflections of the Beyond are of no avail.

6. I believe the omniscient and impersonal Supreme is implacable, beyond concern for human woe, but that there exist lesser divinities,

CONFUCIAN BELIEFS

1. I believe in the presence of the Supreme Ruler in all things, and in Heaven as the Ethical Principle whose law is order, impersonal, and yet interested in humankind.

2. I believe that the purpose of life is to pursue an orderly and reverent existence in accord with "Li" propriety or virtue, so as to become the Superior Being.

3. I believe in the Golden Rule: "Never do to others what you would not like done to you."

4. I believe that Confucius, China's First Sage is the Master of Life whose teachings embody the most profound understanding of earth and Heaven, and Mencius is China's Second Sage.

5. I believe in the writings of Confucius as Scriptural truth and in the Four Sacred Books The Analects; Doctrine of the Mean; Great Learning, and Mencius.

6. I believe that we each have five relationships entailing five duties to our fellow beings, to our ruler; to our father; to our wife; to our elder brother; and to his friend - the foremost being our familial duties.

16

from the high gods who endure for eons to the nature spirits and demons.

7. I believe that all actions create their opposing forces, and the wise will seek inaction in action.

8. I believe that human beings are one of the Ten Thousand Things of manifestation, are finite and will pass; only Tao endures forever.

9. I believe in the oneness of all creation, in the spirituality of the material realms and in universal community.

7. I believe that human nature is inherently good, and evil is an unnatural condition arising from inharmony.

8. I believe that we are rulers of our own lives and fate, free to conduct ourselves as we will and that we should cultivate qualities of benevolence, righteousness, propriety, wisdom and sincerity.

9. I believe the family is the most essential institution among humans and that religion should support the family and the state.

HINDU BELIEFS

1. I believe in one, all-pervasive Supreme Being who is both immanent and transcendent, both Creator and Unmanifest Reality.

2. I believe that the universe undergoes endless cycles of creation, preservation and dissolution.

3. I believe that all souls are evolving toward union with God, will ultimately find Mokaha, spiritual knowledge and liberation from the cycle of rebirth. Not a single soul will be eternally deprived of this destiny.

4. I believe in karma, the law of cause and effect by which individuals create their own destiny by their thoughts, words and deeds.

5. I believe that the soul reincarnates, evolving through many births until all karmas have been resolved.

6. I believe that divine beings exist in unseen inner worlds and that temple worship, rituals and sacraments as well as personal devotionals create a communion with these devas and Gods

7. I believe that a spiritually awakened Sat Guru is essential to know the Transcendent Absolute, as are personal discipline, good conduct, purification, self-inquiry and

BUDDHIST BELIEFS

1. I believe that the Supreme is completely transcendent and can be described as Sunya, a Void or state of non-being.

2. I believe in the Four Noble Truths: 1. that suffering exists; 2.that desire is the cause of suffering; 3. that suffering may be ended by the annihilation of desire; 4.that to end desire one must follow the Eight-Fold Path.

3. I believe in the Eight-Fold Path of right belief, right aims, right speech, right actions, right occupation, right endeavor, right mindfulness and right meditation.

4. I believe that life's aim is to end suffering through the annihilation of individual existence and absorption into Nirvana, the Real.

5. I believe in the "Middle Path", living moderately, avoiding extremes of luxury and asceticism.

6. I believe in the greatness of self-giving love and compassion toward all creatures that live, for these contain merit exceeding the giving of offerings to the gods.

7. I believe in the sanctity of the Buddha and in the sacred scriptures of Buddhism;the Tripitaka (Three

meditation.

8. I believe that all life is sacred and to be loved and revered, and in the practice of ahimsa or non-violence.

9. I believe that no particular religion teaches the only way to salvation above all others, but that all genuine religious paths are facets of God's Pure Love and Light, deserving tolerance and understanding.

Baskets of Wisdom) and/or the Makayana Sutras.

8. I believe that our true nature is divine and eternal, yet our individuality is subject to the change that affects all forms and is therefore transient, dissolving at liberation into Nirvana.

9. I believe in Dharma (the Way), Karma (cause of and effect), Reincarnation, the Sangam(kinship of seekers) and the passage on earth as an opportunity to end the cycle of birth and death.

JAIN BELIEFS

1. I believe in the spiritual lineage of the 24 Tirthankaras ("Ford-Finders') of whom the ascetic sage Mahavira was the last-that they be revered and worshiped above all else.

2. I believe in the sacredness of all life, that one must cease injury to sentient creatures, large and small, and that even unintentional killing creates karma.

3. I believe that God is neither Creator, Father nor Friend. Such human conceptions are limited. And it may be said of God: God is.

4. I believe that each soul is eternal and individual and that we must conquer ourselves by our own efforts and subordinate the worldly to the heavenly to attain Moksha or release.

5. I believe the conquest of oneself can be achieved in ascetic discipline and strict religious observance, that non-ascetics and women have salvation in another life.

6. I believe that the principle governing the succession of life is karma, that our actions both good and bad, bind us and that karma may only be consumed by purification, penance and austerity.

SIKH BELIEFS

1. I believe in God as the Sovereign One, the omnipotent, immortal and personal Creator, a Being beyond time, wno is called Sat Nam for His Name is Truth.

2. I believe that we grow spiritually by living truthfully, serving selflessly and by repetition of the Holy Name and Guru Nanak's Prayer, Japaji.

3. I believe that salvation lies in understanding the Divine Truth and that our surest path lies in faith, love, purity and devotion.

4. I believe in the scriptural and ethical authority of the Adi Granth as God's revelation.

5. I believe that to know God the Guru, is essential as the guide who, absorbed in love of the Real, is able to awaken the soul to its true, Divine nature.

6.I believe in the line of 10 Sikh Gurus: Nanak; Angad; Amardas; Ram Das; Arjan; Har Govind: Har Rai; Tegh Bahadur and Guru Govind Singh- all my teachers.

7. I believe that the world is maya, a vain transitory illusion; and only God is true as all else passes away.

7. I believe in the Agamas and Siddhantas, as the sacred scriptures that guide our moral and spiritual life.

8. I believe in the Three Jewels, right knowledge, right faith and right conduct.

9. I believe the ultimate goal of Mokahais is eternal release from Samsara, the Wheel of Birth and Death and the concomitant attainment of Supreme Knowledge.

8. I believe in adopting the last name "Singh," meaning "Lion" and signifying courage, and in the five symbols: 1. white dress (purity): 2. sword (bravery); 3. iron bracelet (morality); 4.uncut hair and beard (renunciation); and 5.comb (cleanliness).

9. I believe in the natural path and stand opposed to fasting, vegetaranism, pilgrimage, caste, idolatry, celibacy and asceticism.

SHINTO BELIEFS

1. I believe in the "Way of the Gods", Kami-Michl, which asserts nature's sacredness and uniquely reveals the supernatural.

2. I believe there is not a single Supreme Being, but myriad gods, superior beings, among all the wonders of the universe which is not inanimate but filled everywhere with sentient life.

3. I believe in the scriptural authority of the great books known as the Records of Ancient Matters.

4. I believe in the sanctity of cleanliness and purity - of body and spirit - and that impurity is a religious transgression.

5. I believe that the State is a divine institution whose laws should not be transgressed and to which individuals must sacrifice their own needs.

6. I believe in the moral and spiritual uprightness as the cornerstone of religious ethics and in the supreme value of loyalty.

7. I believe that the supernatural reveals itself through all that is natural and beautiful, and value these above philosophical or theological doctrine.

ZOROASTRIAN BELIEFS

1. I believe there are two Great Beings in the universe. One created people and all that is good, beautiful and true, while the other vivifies all that is evil, ugly and destructive.

2. I believe that people have free will to align themselves with good or evil, and when all humankind is in harmony with the God of good, Evil will be conquered.

3. I believe the soul is immortal and upon death crosses over Hell by a Narrow Bridge - the good crossing safely to heaven and the evil falling into Hell.

4. I believe that a savior will appear at the end of time, born of a virgin, reviving the dead, rewarding the good and punishing the evil and thereafter Good will reign.

5. I believe that Zoroaster, also known as Zarathustra, is the foremost Prophet of God.

6. I believe in the scriptural authority of the Zend Aveda.

7. I believe that purity is the first virtue, truth the second and charity the third - and that people must discipline themselves by good thought, words and deeds.

8. I believe marriage excels continence, action excels

8. I believe that whatever is, is divine Spirit, that the world is one community, that all people are capable of deep affinity with the Divine and there exists no evil in the world whatsoever.

9. I believe in the practical use of ceremony and ritual, and in the worship of the deities that animate nature, including the Sun Goddess, Star God and Storm God.

contemplation, forgiveness excels revenge.

9. I believe in God as seven persons: Eternal Light; Right and Justice; Goodness and Love; Strength of Spirit; Piety and Faith; Health and Perfection; and Immortality - and that God may best be worshiped through the representation of fire.

JUDAIC BELIEFS

1. I believe in the one God and Creator who is incorporeal and transcendent, who cares for the world and its creatures, rewarding the good and punishing the evil.

2. I believe in the Prophets, of which Moses was God's foremost, and that the commandments revealed to him by God on Mount Sinai are human's highest law.

3. I believe in the Torah as God's word and scripture, the five Old Testament books. They are God's only immutable law.

4. I believe that upon death the soul goes to Heaven (or to Hell first if sinful), that one day the Messiah will appear on earth. There will be a Day of Judgment, and the dead shall physically arise to life everlasting.

5. I believe that the universe is not eternal but was created and will be destroyed by God.

6. I believe that no priest should intervene in the relationship of human and God, nor should God be represented in any form, nor should any being be worshiped other than the One God, Yahweh.

7. I believe in human's spiritualization through adherence to the law, justice, charity and honesty.

CHRISTIAN BELIEFS

1. I believe in God the Father, Creator of the universe, reigning forever distinct over humankind, his beloved creation.

2. I believe humans are born sinners, and that they may know salvation only through the Savior, Jesus Christ, God's only begotten Son.

3. I believe that Jesus Christ was born of Mary, a virgin.

4. I believe that Jesus Christ was crucified on the cross, then resurrected from the dead and now sits at the right hand of the Father as the final judge of the dead, and that He will return again as prophesied.

5. I believe that the soul is embodied for a single lifetime, but is immortal and accountable to God for all thoughts and actions.

6. I believe in the historical truth of the Holy Bible, that it is sacred scripture of the highest authority and the only word of God.

7. I believe that upon death, the soul enters heaven, Purgatory or Hell, awaiting the Last Judgement when the dead shall rise again, to enjoy life everlasting and the unsaved to suffer eternally.

8. I believe in the intrinsic goodness of humankind and the affirmative

8. I believe that God has established a unique spiritual covenant with the Hebrew people to uphold for humankind the highest standards of monotheism and piety.

9. I believe in the duty of the family to make the home a House of God through devotions and ritual, prayers, sacred festivals and observation of the Holy Days.

nature of life and in the priceless value of love, charity and faith.

9. I believe in the Holy Trinity of God who reveals Himself as Father, Son and Holy Ghost, and in the existence of Satan, the personification of evil.

ISLAMIC BELIEFS

1. I believe that Allah is the Supreme Creator and Sustainer, all-knowing and transcendent and yet the arbiter of good and evil, the final judge of people.

2. I believe in the Five Pillars of Faith: 1. Praying five times daily, 2. Charity through alms giving, 3. Fasting during the ninth month, 4. Pilgrimage to Holy Mecca, 5. Profession of faith by acknowledging "There is no God but Allah, Mohammed is His Prophet".

3. I believe in the Koran as the Word of God and sacred scripture mediated through the Angel Gabriel to Mohammed.

4. I believe in the direct communion of each person with God, that all are equal in the eyes of God and therefore priests or other intercessors are unneeded.

5. I believe in the pure transcendence of God, great beyond imagining - no form or idol can be worshiped in God's Name.

6. I believe that the soul of people is Immortal, embodied once on earth then entering Heaven or Hell upon death according to its conduct and faith on earth.

7. I believe in the Last Judgment and that people should stand in

FAITHS' BELIEFS

1. I believe in the fundamental unity and common Source of all religions (Baha'i).

2. I believe humans' natural spirituality is best expressed in loving and practical aid to fellow humans, rather than metaphysical inquiry. (Humanitarianism).

3. I believe in the unity of religions, the efficacy of devotion, and service and in the Living Incarnation of the Divine. (Salam).

4. I believe that spiritual progress comes through analysis of current and past life experiences which resolve past karma most directly. (Scientology).

5. I believe that there is no God beyond the Divine within humans and no truth beyond existential freedom, that all religions imprison humans, causing repression, fear and poverty. (Rajneeshism).

6. I believe humans' sense of the sacred can be fulfilled naturally, without formal worship, houses of God, ceremony, creeds or theology. (Various faiths).

7. I believe religion consists of unitive and direct mystical experience which should be the objective of every religious aspirant. (Mysticism).

humble awe and fear of God's wrathful and vengeful power.

8. I believe that truthfulness should be observed in all circumstances, even though it may bring injury or pain.

9. I believe that salvation is only obtained through God's Grace and humans should do good and avoid all sins, especially drunkenness, usury and gambling.

8. I believe that the cultivation of occult powers including ESP, astral travel, past life readings, etc., is the highest pursuit of that which is spiritual. (Occultism).

9. I believe there is no knowable providential order, that death is permanent, that God does not exist and that the highest life is one of intense consciousness and free, creative action. (Existentialism).

DECLARATION OF INTERDEPENDENCE

We the People hereby declare our interdependence - our connection to the Source of All Life and to all life forms. We affirm that diverse individuals, groups, and networks are necessary for the creative development of humanity; and that to strengthen UNITY-AND-DIVERSITY throughout the universe is our individual responsibility and privilege.

We therefore pledge -

* To affirm the existence of a Supreme Beingness, called by any name or no name;
* To advance both individual initiative and human fellowship through mutual trust, understanding, and respect;
* To seek the truth in the spirit of love;
* To integrate reason and faith, science and religion;
* To ensure that all aspects of life be kept in dynamic balance for maximum health and well-being;
* To respect the teachings of the prophets and sages of all times and cultures.
* To provide present and future generations with the opportunity for full realization of their potential; and
* To build with joy a new civilization of freedom, justice, and peace founded on *reverence for life*.

We the People therefore proclaim our interdependence. We shall kindle the torch of hope, link hands over space and time, and fulfill our interdependence through action.

Unity-and-Diversity World Council

A UNIVERSAL DECLARATION OF MORAL AND SPIRITUAL VALUES

If we seek real fulfillment in our lives, following a particular path and yet striving to realize our total potential and to relate to all paths, we will find some place for each of the following twelve guidelines. Let us - -

(1) Be in tune with the Spirit of All Life, called by any name or no name;

(2) Practice meditation, contemplation, and prayer;

(3) Show all-embracing love toward all beings;

(4) Experience the true nature of our self and our universe;

(5) Cultivate truth, goodness, beauty, and respect;

(6) Live simply and harmoniously with our whole self;

(7) Use our energy for vigorous and constructive activity;

(8) Rejoice in our connection with all human beings and all life;

(9) Strive for peaceful family and community development;

(10) Get involved in improving the world's condition;

(11) Preserve the best of our universal heritage; and

(12) Take heart and act upon our ideals;

Let us also seek to experience each of these dimensions of our total being and discover their interrelationships. At the same time, may we endeavor to become responsible participants in the emerging global civilization based on the dynamic integration of diversity among all peoples and all life.

EXPANDING ON THESE GUIDELINES

(1) SPIRIT OF ALL LIFE - Let ourselves be used for the growth of Spirit in all peoples and all life. Our Inner Knowing is to be trusted; by being attuned to it we find security. Through faith in the Supreme Beingness, called by any name or no name, comes a feeling of hope and thankfulness which can be passed on to all.

(2) MEDITATION, CONTEMPLATION, AND PRAYER - Let us set aside time for going within as a necessary part of our existence. Its fruits are dispassion, inner peace, understanding, strength of conviction, and union with the Ultimate. Through this reflection let us seek to transform our lives.

(3) ALL EMBRACING LOVE - From compassionate concern for others come love, humility, mercy, nonviolence, and a quest for world unification. Let us treat others as we would have others treat us.

(4) NATURE OF SELF AND UNIVERSE - Let us be open to experiences in life which come unsought. May we allow our true nature to unfold as it will, not trying to dominate or control where patience and letting alone are the better way.

(5) TRUTH, GOODNESS, BEAUTY, AND RESPECT - Let us cultivate these qualities in our everyday existence.

(6) SIMPLICITY AND HARMONY - Let us seek to live a simple and balanced life. May we be moderate in our desires and sincere in our attitude toward others. Let us relax, be calm and unassuming.

(7) VIGOROUS AND CONSTRUCTIVE ACTIVITY - Let us express the dynamic of our faith by summoning our utmost vitality in its behalf. Let us seek creative work to meet human needs.

(8) REJOICING AND CONNECTING - Let us celebrate all human beings and all forms of life. May we accept the challenge of living and show gratitude for the meaning in each moment of existence.

(9) FAMILY AND COMMUNITY - Let us participate in family and community development with enthusiasm and common sense.

(10) WORLD IMPROVEMENT - Let us take part in the improvement of the world's condition. May we apply appropriate scientific techniques to the solution of global problems.

(11) PRESERVATION OF THE BEST - Let us develop awareness, appreciation, and concern to preserve the best of our total heritage.

(12) ACTING UPON OUR IDEALS - Let us complete the cycle of our own life by fulfilling our ideals and dreams. May we act in concert with one another to improve the quality of our lives, our institutions, our environment, and our global future.

Unity-and-Diversity World Council

A GLOBAL ETHIC
1993 PARLIAMENT OF THE WORLD'S RELIGIONS

We Declare:

We are interdependent. Each of us depends on the well-being of the whole, and so we have respect for the community of living beings, for people, animals, and plants, and for the preservation of the Earth, the air, water and soil.

We take individual responsibility for all we do. All our decisions, actions, and failure to act have consequences.

We must treat others as we wish others to treat us. We make commitment to respect life and dignity, individuality and diversity, so that every person is treated humanely, without exception. We must have patience and acceptance. We must be able to forgive, learning from the past but never allowing ourselves to be enslaved by memories of hate. Opening our hearts to one another, we must sink our narrow differences for the cause of the world community, practicing a culture of solidarity and relatedness.

We consider humankind our family. We must strive to be kind and generous. We must not live for ourselves alone, but should also serve others, never forgetting the children, the aged, the poor, the suffering, the disabled, the refugees, and the lonely. No person should ever be considered or treated as a second-class citizen, or be exploited in any way whatsoever. There should be equal partnership between men and women. We must not commit any kind of sexual immorality . We must put behind us all forms of domination or abuse.

We commit ourselves to a culture of non-violence, respect, justice, and peace. We shall not oppress, injure, torture, or kill other human beings, forsaking violence as a means of settling differences.

We must strive for a just social and economic order, in which everyone has an equal chance to reach full potential as a human being. We must speak and act truthfully and with compassion, dealing fairly with all, and avoiding prejudice and hatred. We must not steal. We must move beyond the dominance of greed for power, prestige, money, and consumption to make a just and peaceful world.

Earth cannot be changed for the better unless the consciousness of individuals is changed first. We pledge to increase our awareness of disciplining our minds, by meditation, by prayer, or by positive thinking. Without risk and a readiness to sacrifice there can be no fundamental change in our situation. Therefore we commit ourselves to this global ethic, to understanding one another, and to socially-beneficial, peace-fostering, and nature-friendly ways of life.

We invite all people, whether religious or not, to do the same.

The Principles of a Global Ethic

Our world is experiencing a fundamental crisis: A crisis in global economy, global ecology, and global politics. The lack of a grand vision, the tangle of unresolved problems, political paralysis, mediocre political leadership with little insight or foresight, and in general too little sense for the commonweal are seen everywhere: Too many old answers to new challenges.

Hundreds of millions of human beings on our planet increasingly suffer from unemployment, poverty, hunger, and the destruction of their families. Hope for a lasting peace among nations slips away from us. There are tensions between the sexes and generations. Children die, kill, and are killed. More and more countries are shaken by corruption in politics and business. It is increasingly difficult to live together peacefully in our cities because of social, racial, and ethnic conflicts, the abuse of drugs, organized crime and even anarchy. Even neighbors often live in fear of one another. Our planet continues to be ruthlessly plundered. A collapse of the ecosystem threatens us.

Time and again we see leaders and members of religions incite aggression, fanaticism, hate, and zenophobia - even inspire and legitimate violent and bloody conflicts. Religion often is misused for purely power-political goals, including war. We are filled with disgust.

We condemn these blights and declare that they need not be. An ethic already exists within the religious teachings of the world which can counter the global distress. Of course this ethic provides no direct solution for all the immense problems of the world, but it does supply the moral foundation for a better individual and global order: A vision which can lead women and men away from despair, and society away from chaos.

We are persons who have committed ourselves to the precepts and practices of the world's religions. We confirm that there is already a consensus among the religions which can be the basis for a global ethic - a minimal *fundamental consensus* concerning binding *values*, irrevocable *standards*, and *fundamental moral attitudes*.

1. No new global order without a new global ethic!

We women and men of various religions and regions of Earth therefore address all people, religious and non-religious. We wish to express the following convictions which we hold in common:

* We all have a responsibility for a better global order.

* Our involvement for the sake of human rights, freedom, justice, peace, and the preservation of Earth is absolutely necessary.

* Our different religious and cultural traditions must not prevent our common involvement in opposing all forms of inhumanity and working for greater humaneness.

* The principles expressed in this Global Ethic can be affirmed by all persons with ethical convictions, whether religiously grounded or not.

* As religious and spiritual persons we base our lives on an Ultimate Reality, and draw spiritual power and hope therefrom, in trust, in prayer or meditation, in word or silence. We have a special responsibility for the welfare of all humanity and care for the planet Earth. We do not consider ourselves better than other women and men, but we trust that the ancient wisdom of our religions can point the way for the future.

After two world wars and the end of the cold war, the collapse of fascism and nazism, the shaking to the foundations of communism and colonialism, humanity has entered a new phase of its history. Today we possess sufficient economic, cultural, and spiritual resources to introduce a better global order. But old and new ethnic, national, social, economic, and religious tensions threaten the peaceful building of a better world. We have experienced greater technological progress than ever before, yet we see that world-wide poverty, hunger, death of children, unemployment, misery, and the destruction of nature have not diminished but rather have increased.. Many peoples are threatened with economic ruin, social disarray, political marginalization, ecological catastrophe, and national collapse.

In such a dramatic global situation humanity needs a vision of peoples living peacefully together, of ethnic and ethical groupings and of religions sharing responsibility for the care of Earth. A vision rests on hopes, goals ideals, standards. But all over the world these have slipped from our hands. Yet we are convinced that, despite their frequent abuses and failure, it is the communities of faith who bear a responsibility to demonstrate that such hopes, ideals, and standards can be guarded, grounded, and lived. This is especially true in the modern state. Guarantees of freedom of conscience and religion are necessary but they do not substitute for binding values, convictions, and norms which are valid for all humans regardless of their social origin, sex, skin color, language, or religion.

We are convinced of the fundamental unity of the human family on Earth. We recall the 1948 Universal Declaration of Human Rights of the United Nations. What it formally proclaimed on the level of rights we wish to confirm and deepen here from the perspective of an ethic: The full realization of the intrinsic dignity of the human person, the inalienable freedom and quality in principle of all humans, and the necessary solidarity and interdependence of all humans with each other.

On the basis of personal experiences and the burdensome history of our planet we have learned

* that a better global order cannot be created or enforced by laws, prescriptions, and conventions alone;

* that the realization of peace, justice, and the protection of Earth depends on the insight and readiness of men and women to act justly;

* that action in favor of rights and freedoms presumes a consciousness of responsibility and duty, and that therefore both the minds and hearts of women and men must be addressed;

35

* that rights without morality cannot long endure, and that *there will be no better global order without a global ethic.*

By a global ethic we do not mean a global ideology or a single unified religion beyond all existing religions, and certainly not the domination of one religion over all others. By a global ethic we mean a fundamental consensus on binding values, irrevocable standards, and personal attitudes. Without such a fundamental consensus on an ethic, sooner or later every community will be threatened by chaos or dictatorship, and individuals will despair.

II. A fundamental demand: Every human being must be treated humanely.

We all are fallible, imperfect men and women with limitations and defects. We know the reality of evil. Precisely because of this, we feel compelled for the sake of global welfare to express what the fundamental elements of a global ethic should be - for individuals as well as for communities and organizations, for states as well as for the religions themselves. We trust that our often millennia-old religious and ethical traditions provide an ethic which is convincing and practicable for all women and men of good will, religious and non-religious.

At the same time we know that our various religious and ethical traditions often offer very different bases for what is helpful and what is unhelpful for men and women, what is right and what is wrong, what is good and what is evil. We do not wish to gloss over or ignore the serious differences among the individual religions. However, they should not hinder us from proclaiming publicly those things which we already hold in common and which we jointly affirm, each on the basis of our own religious or ethical grounds.

We know that religions cannot solve the environmental, economic, political, and social problems of Earth. However they can provide what obviously cannot be attained by economic plans, political programs, or legal regulations alone: A change in the inner orientation, the whole mentality, the "hearts" of people, and a conversion from a false path to a new orientation for life. Humankind urgently needs social and ecological reforms, but it needs spiritual renewal just as urgently. As religious or spiritual persons we commit ourselves to this task. The spiritual powers of the religions can offer a fundamental sense of trust, a ground of meaning, ultimate standards, and a spiritual home. Of course religions are credible only when they eliminate those conflicts which spring from the religions themselves, dismantling mutual arrogance, mistrust, prejudice, and even hostile images, and thus demonstrate respect for the traditions, holy places, feasts, and rituals of people who believe differently.

Now as before, women and men are treated inhumanly all over the world. They are robbed of their opportunities and their freedom; their human rights are trampled underfoot; their dignity is disregarded. But might does not make right! In the face of all inhumanity our religious and ethical convictions demand that *every human being must be treated humanely!*

This means that every human being without distinction of age, sex, race, skin color, physical or mental ability, language, religion, political view, or national or social origin possesses an inalienable and untouchable dignity, and everyone, the individual as well as the state, is therefore obliged to honor this dignity and protect it. Humans must always be the subjects of rights, must be ends, never mere means, never objects of commercialization and industrialization in economics, politics and media, in research institutes, and industrial corporations. No one stands "above good and evil" - no human being, no social class, no influential interest group, no cartel, no police apparatus, no army, and no state. On the contrary: Possessed of reason and conscience, every human is obliged to behave in a genuinely human fashion, to do good and avoid evil!

It is the intention of this Global Ethic to clarify what this means. In it we wish to recall irrevocable, unconditional ethical norms. These should not be bonds and chains, but helps and supports for people to find and realize once again their lives' direction, values, orientations, and meaning.

There is a principle which is found and has persisted in many religious and ethical traditions of humankind for thousands of years: *What you do not wish done to yourself, do not do to others.* Or in positive terms: *What you wish done to yourself, do to others!*

Every form of egoism should be rejected: All selfishness, whether individual or collective, whether in the form of class thinking, racism, nationalism, or sexism. We condemn these because they prevent humans from being authentically human. Self-determination and self-realization are thoroughly legitimate so long as they are not separated from human self-responsibility and global responsibility, that is, from responsibility for fellow humans and for the planet Earth.

This principle implies very concrete standards to which we humans should hold firm. From it arise four broad, ancient guidelines for human behavior which are found in most of the religions of the world.

III.   Irrevocable directives.

1. Commitment to a Culture of Non-violence and Respect for Life.

Numberless women and men of all regions and religions strive to lead lives not determined by egoism but by commitment to their fellow humans and to the world around them. Nevertheless, all over the world we find endless hatred, envy, jealousy, and violence, not only between individuals but also between social and ethnic groups, between classes, races, nations, and religions. The use of violence, drug trafficking and organized crime, often equipped with new technical possibilities, has reached global proportions. Many places still are ruled by terror "from above;" dictators oppress their own people, and institutional violence is widespread. Even in some countries where laws exist to protect individual freedoms, prisoners are tortured, men and women are mutilated, hostages are killed.

a) In the great ancient religious and ethical traditions of humankind we find the directive: *You shall not kill!* Or in positive terms: *Have respect for life!* Let us reflect anew on the consequences of this ancient directive: All people have a right to life, safety, and the free development of personality insofar as they do not injure the rights of others. No one has the right physically or psychically to torture, injure, much less kill, any other human being. And no people, no state, no race, no religion has the right to hate, to discriminate against, to "cleanse," to exile, much less to liquidate a "foreign" minority which is different in behavior or holds different beliefs.

b) Of course, wherever there are humans there will be conflicts. Such conflicts, however, should be resolved without violence within a framework of justice. This is true for states as well as for individuals. Persons who hold political power must work within the framework of a just order and commit themselves to the most non-violent, peaceful solutions possible. And they should work for this within an international order of peace which itself has need of protection and defense against perpetrators of violence.

Armament is a mistaken path; disarmament is the commandment of the times. Let no one be deceived: There is no survival for humanity without global peace!

c) Young people must learn at home and in school that violence may not be a means of settling differences with others. Only thus can a culture of non-violence be created.

d) A human person is infinitely precious and must be unconditionally protected. But likewise the lives of animals and plants which inhabit this planet with us deserve protection, preservation, and care. Limitless exploitation of the natural foundations of life, ruthless destruction of the biosphere, and militarization of the cosmos are all outrages. As human beings we have a special responsibility - especially with a view to future generations - for Earth and the cosmos, for the air, water, and soil. We are all intertwined together in this cosmos and we are all dependent on each other. Each one of us depends on the welfare of all. Therefore the dominance of humanity over nature and the cosmos must not be encouraged. Instead we must cultivate living in harmony with nature and the cosmos.

e) To be authentically human in the spirit of our great religious and ethical traditions means that in public as well as in private life we must be concerned for others and ready to help. We must never be ruthless and brutal. Every people, every race, every religion must show tolerance and respect - indeed high appreciation - for every other. Minorities need protection and support, whether they be racial, ethnic, or religious.

2. Commitment to a Culture of Solidarity and a Just Economic Order.

Numberless men and women of all regions and religions strive to live their lives in solidarity with one another and to work for authentic fulfillment of their vocations. Nevertheless, all over the world we find endless hunger, deficiency, and need. Not only individuals, but especially unjust institutions and structures are responsible for these tragedies. Millions of people are without work; millions are exploited by poor wages, forced to the edges of society, with their possibilities for the future destroyed. In many lands the gap between the poor and the rich, between the powerful and the powerless is immense. We live in a world in which totalitarian state socialism as well as unbridled capitalism have hollowed out and destroyed many ethical and spiritual values. A materialistic mentality breeds greed for unlimited profit and a grasping for endless plunder. These demands claim more and more of the community's resources without obliging the individual to contribute more. The cancerous social evil of corruption thrives in the developing countries and in the developed countries alike.

a) In the great ancient religious and ethical traditions of humankind we find the directive: *You shall not steal!* Or in positive terms: *Deal honestly and fairly!* Let us reflect anew on the consequences of this ancient directive: No one has the right to rob or dispossess in any way whatsoever any other person or the commonweal. Further, no one has the right to use her or his possessions without concern for the needs of society and Earth.

b) Where extreme poverty reigns, helplessness and despair spread, and theft occurs again and again for the sake of survival. Where power and wealth are accumulated ruthlessly, feelings of envy, resentment, and deadly hatred and rebellion inevitably well up in the disadvantaged and marginalized. This leads to a vicious circle of violence and counter-violence. Let no one be deceived: There is no global peace without global justice!

c) Young people must learn at home and in school that property, limited though it may be, carries with it an obligation, and that its uses should at the same time serve the common good. Only thus can a just economic order be built up.

d) If the plight of the poorest billions of humans on this planet, particularly women and children, is to be improved, the world economy must be structured more justly. Individual good deeds, and assistance projects, indispensable though they be, are insufficient. The participation of all states and the authority of international organizations are needed to build just economic institutions.

A solution which can be supported by all sides must be sought for the debt crisis and the poverty of the dissolving second world, and even more the third world. Of course conflicts of interest are unavoidable. In the developed countries, a distinction must be made between necessary and limitless consumption, between socially beneficial and non-beneficial uses of property, between justified and unjustified uses of natural resources, and between a profit-only and a socially beneficial and ecologically oriented market economy. Even the developing nations must search their national consciences.

Wherever those ruling threaten to repress those ruled, wherever institutions threaten persons, and wherever might oppresses right, we are obligated to resist - whenever possible non-violently.

e) To be authentically human in the spirit of our great religious and ethical traditions means the following:

* We must utilize economic and political power for service to humanity instead of misusing it in ruthless battles for domination. We must develop a spirit of compassion with those who suffer, with special care for the children, the aged, the poor, the disabled, the refugees, and the lonely.

* We must cultivate mutual respect and consideration, so as to reach a reasonable balance of interests, instead of thinking only of unlimited power and unavoidable competitive struggles.

* We must value a sense of moderation and modesty instead of an unquenchable greed for money, prestige, and consumption. In greed humans lose their "souls," their freedom, their composure, their inner peace, and thus that which makes them human.

3. Commitment to a Culture of Tolerance and a Life of Truthfulness.

Numberless women and men of all regions and religions strive to lead lives of honesty and truthfulness. Nevertheless, all over the world we find endless lies and deceit, swindling and hypocrisy, ideology and demagoguery:

* Politicians and business people who use lies as a means to success;

* Mass media which spread ideological propaganda instead of accurate reporting, misinformation instead of information, cynical commercial interest instead of loyalty to the truth;

* Scientists and researchers who give themselves over to morally questionable ideological or political programs or to economic interest groups, or who justify research which violates fundamental ethical values;

* Representatives of religions who dismiss other religions as of little value and who preach fanaticism and intolerance instead of respect and understanding.

a) In the great ancient religious and ethical traditions of humankind we find the directive: *You shall not lie*! Or in positive terms: *Speak and act truthfully!* Let us reflect anew on the consequences of this ancient directive: No woman or man, no institution, no state or church or religious community has the right to speak lies to other humans.

b) This is especially true
* for those who work in the mass media, to whom we entrust the freedom to report for the sake of truth and to whom we thus grant the office of guardian. They do not stand above morality but have the obligation to respect human dignity, human rights, and fundamental values. They are duty-bound to objectivity, fairness, and the preservation of human dignity. They have no right to intrude into individuals' private spheres, to manipulate public opinion, or to distort reality.

* for artists, writers, and scientists, to whom we entrust artistic and academic freedom. They are not exempt from general ethical standards and must serve the truth;

* for the leaders of countries, politicians, and political parties, to whom we entrust our own freedoms. When they lie in the faces of their people, when they manipulate the truth, or when they are guilty of venality or ruthlessness in domestic or foreign affairs, they forsake their credibility and deserve to lose their offices and their voters. Conversely, public opinion should support those politicians who dare to speak the truth to the people at all times.

* finally, for representatives of religion. When they stir up prejudice, hatred, and enmity towards those of different belief, or even incite or legitimate religious wars, they deserve the condemnation of humankind and the loss of their adherents.

Let no one be deceived: There is no global justice without truthfulness and humaneness!

c) Young people must learn at home and in school to think, speak, and act truthfully. They have a right to information and education to be able to make the decisions that will form their lives. Without an ethical formation they will hardly be able to distinguish the important from the unimportant. In the daily flood of information, ethical standards will help them discern when opinions are portrayed as facts, interests veiled, tendencies exaggerated, and facts twisted.

d) To be authentically human in the spirit of our great religious and ethical traditions means the following:

* We must not confuse freedom with arbitrariness or pluralism with indifference to truth. We must cultivate truthfulness in all our relationships instead of dishonesty, dissembling, and opportunism.

* We must constantly seek truth and incorruptible sincerity instead of spreading ideological or partisan half-truths.

* We must courageously serve the truth and we must remain constant and trustworthy, instead of yielding to opportunistic accommodation to life.

4. Commitment to a Culture of Equal Rights and Partnership Between Men and Women.

Numberless men and women of all regions and religions strive to live their lives in a spirit of partnership and responsible action in the areas of love, sexuality, and family. Nevertheless, all over the world there are condemnable forms of patriarchy, domination of one sex over the other, exploitation of women, sexual misuse of children, and forced prostitution. Too frequently, social inequities force women and even children into prostitution as a means of survival - particularly in less developed countries.

a) In the great ancient religious and ethical traditions of humankind we find the directive: *You shall not commit sexual immorality!* Or in positive terms: *Respect and love one another!* Let us reflect anew on the consequences of this ancient directive: No one has the right to degrade others to mere sex objects, to lead them into or hold them in sexual dependency.

b) We condemn sexual exploitation and sexual discrimination as one of the worst forms of human degradation. We have the duty to resist wherever the domination of one sex over the other is preached - even in the name of religious conviction; wherever sexual exploitation is tolerated, wherever prostitution is fostered or children are misused. Let no one be deceived: There is no authentic humaneness without a living together in partnership!

c) Young people must learn at home and in school that sexuality is not a negative, destructive, or exploitative force, but creative and affirmative. Sexuality as a life-affirming shaper of community can only be effective when partners accept the responsibilities of caring for one another's happiness.

d) The relationship between women and men should be characterized not by patronizing behavior or exploitation, but by love, partnership, and trustworthiness. Human fulfillment is not identical with sexual pleasure. Sexuality should express and reinforce a loving relationship lived by equal partners. Some religious traditions know the ideal of a voluntary renunciation of the full use of sexuality. Voluntary renunciation also can be an expression of identity and meaningful fulfillment.

e) The social institution of marriage, despite all its cultural and religious variety, is characterized by love, loyalty, and permanence. It aims at and should guarantee security and mutual support to husband, wife, and child. It should secure the rights of all family members.

All lands and cultures should develop economic and social relationships which will enable marriage and family life worthy of human beings, especially for older people. Children have a right of access to education. Parents should not exploit children, nor children parents. Their relationships should reflect mutual respect, appreciation, and concern.

f) To be authentically human in the spirit of our great religious and ethical traditions means the following:

* We need mutual respect, partnership, and understanding, instead of patriarchal domination and degradation, which are expressions of violence and engender counter-violence.

* We need mutual concern, tolerance, readiness for reconciliation, and love, instead of any form of possessive lust or sexual misuse.

Only what has already been experienced in personal and familial relationships can be practiced on the level of nations and religions.

IV. *A Transformation of Consciousness!*

Historical experience demonstrates the following: Earth cannot be changed for the better unless we achieve a transformation in the consciousness of individuals and in public life. The possibilities for transformation have already been glimpsed in areas such as war and peace, economy, and ecology, where in recent decades fundamental changes have taken place. This transformation must also be achieved in the area of ethics and values!

Every individual has intrinsic dignity and inalienable rights, and each also has an inescapable responsibility for what she or he does and does not do. All our decisions and deeds, even our omissions and failures, have consequences.

Keeping this sense of responsibility alive, defending it and passing it on to future generations is the special task of religions.

We are realistic about what we have achieved in this consensus, and so we urge that the following be observed:

1. A universal consensus on many disputed ethical questions (from bio- and sexual ethics through mass media and scientific ethics to economic and political ethics) will be difficult to attain. Nevertheless, even for many controversial questions suitable solutions should be attainable in the spirit of the fundamental principles we have jointly developed here.

2. In many areas of life a new consciousness of ethical responsibility has already arisen. Therefore we would be pleased if as many professions as possible such as those of physicians, scientists, business people, journalists, and politicians, would develop up-to-date codes of ethics which would provide specific guidelines for the vexing questions of these particular professions.

3. Above all, we urge the various communities of faith to formulate their very specific ethics: What does each faith tradition have to say, for example, about the meaning of life and death, the enduring of suffering and the forgiveness of guilt, about selfless sacrifice and the necessity of

renunciation, about compassion and joy. These will deepen, and make more specific, the already discernible global ethic.

In conclusion, we appeal to all the inhabitants of this planet. Earth cannot be changed for the better unless the consciousness of individuals is changed. We pledge to work for such transformation in individual and collective consciousness, for the awakening of our spiritual powers through reflection, meditation, prayer, or positive thinking, for a conversion of the heart. Together we can move mountains! Without a willingness to take risks and a readiness to sacrifice there can be no fundamental change in our situation! Therefore we commit ourselves to a common global ethic, to better mutual understanding, as well as to socially-beneficial, peace-fostering, and Earth-friendly ways of life.

We invite all men and women, whether religious or not, to do the same.

UNIVERSAL DECLARATION OF HUMAN RIGHTS

(incorporating gender equality)

On December 10, 1948 the General Assembly of the United Nations adopted and proclaimed the Universal Declaration of Human Rights the full text of which appears in the following pages. Following this historic act the Assembly called upon all Member countries to publicize the text of the Declaration and "to cause it to be disseminated, displayed, read and expounded principally in schools and other educational institutions, without distinction based on the political status of countries or territories."

PREAMBLE

Whereas recognition of the inherent dignity and of the equal and inalienable **rights** of all members of the **human** family is the foundation of freedom, justice and peace in the world,

Whereas disregard and contempt for **human rights** have resulted in barbarous acts which have outraged the conscience of humankind, and the advent of a world in which **human beings** shall enjoy freedom of speech and belief and freedom from fear and want has been proclaimed as the highest aspiration of the common people,

Whereas it is essential, if humans not to be compelled to have recourse, as a last resort, to rebellion against tyranny and oppression, that **human rights** should be protected by the rule of law,

Whereas it is essential to promote the development of friendly relations between nations,

Whereas the peoples of the United Nations have in the Charter reaffirmed their faith in fundamental **human rights**, in the dignity and worth of the **human** person and in the equal **rights** of men and women and have determined to promote social progress and better standards of life in larger freedom,

Whereas Member States have pledged themselves to achieve, in co-operation with the United Nations, the promotion of universal respect for and observance of **human rights** and fundamental freedoms,

Whereas a common understanding of these **rights** and freedoms is the greatest importance for the full realization of this pledge,

Now, Therefore THE GENERAL ASSEMBLY proclaims THIS UNIVERSAL DECLARATION OF HUMAN RIGHTS as a common standard of achievement for all peoples and all nations, to the end that every individual and every organ of society, keeping this **Declaration** constantly in mind, shall strive by teaching and education to promote respect for these **rights** and freedoms and by progressive measures, national and international, to secure their universal and effective recognition and observance, both among the peoples of Member States themselves and among the peoples of territories under their jurisdiction.

Article 1.

All **human** beings are born free and equal in dignity and **rights**. They are endowed with reason and conscience and should act towards one another in a spirit of kinship.

Article 2.

Everyone is entitled to all the **rights** and freedoms set forth in this **Declaration**, without distinction of any kind, such as race, color, sex, language, religion, political or other opinion, national or social origin, property, birth or other status. Furthermore, no distinction shall be made on the basis of the political, jurisdictional or international status of the country or territory to which a person belongs, whether it be independent, trust, non-self-governing or under any other limitation of sovereignty.

Article 3.

Everyone has the right to life, liberty and security of person.

Article 4.

No one shall be held in slavery or servitude; slavery and the slave trade shall be prohibited in all their forms.

Article 5.

No one shall be subjected to torture or to cruel, inhuman or degrading treatment or punishment.

Article 6.

Everyone has the right to recognition everywhere as a person before the law.

Article 7.

All are equal before the law and are entitled without any discrimination to equal protection of the law. All are entitled to equal protection against any discrimination in violation of this **Declaration** and against any incitement to such discrimination.

Article 8.

All persons have the right to an effective remedy by the competent national tribunals for acts violating the fundamental **rights** granted them by the constitution or by law.

Article 9.

No one shall be subjected to arbitrary arrest, detention or exile.

Article 10.

All persons are entitled in full equality to a fair and public hearing by an independent and impartial tribunal, in the determination of their **rights** and obligations and of any criminal charge against them.

Article 11.

(1) All persons charged with a penal offence have the right to be presumed innocent until proved guilty according to law in a public trial at which they have had all the guarantees necessary for their defence.

(2) No one shall be held guilty of any penal offence on account of any act or omission which did not constitute a penal offence, under national or international law, at the time when it was committed. Nor shall a heavier penalty be imposed than the one that was applicable at the time the penal offence was committed.

Article 12.

No persons shall be subjected to arbitrary interference with their privacy, family, home or correspondence, nor to attacks upon their honor and reputation. Everyone has the right to the protection of the law against such interference or attacks.

Article 13.

(1) Everyone has the right to freedom of movement and residence within the borders of each state.

(2) All persons have the right to leave any country, including their own, and to return to their country.

Article 14.

(1) Everyone has the right to seek and enjoy in other countries asylum from persecution.

(2) This right may not be invoked in the case of prosecutions genuinely arising from non-political crimes or from acts contrary to the purposes and principles of the United Nations.

Article 15.

(1) Everyone has the right to a nationality.

(2) No persons shall be arbitrarily deprived of their nationality nor denied the right to change their nationality.

Article 16.

(1) Men and women of full age, without any limitation due to race, nationality or religion, have the right to marry and to found a family. They are entitled to equal **rights** as to marriage, during marriage and at its dissolution.

(2) Marriage shall be entered into only with the free and full consent of the intending spouses.

(3) The family is the natural and fundamental group unit of society and is entitled to protection by society and the State.

Article 17.

(1) Everyone has the right to own property alone as well as in association with others.

(2) No persons shall be arbitrarily deprived of their property.

Article 18.

Everyone has the right to freedom of thought, conscience and religion; this right includes freedom to change their religion or belief, and freedom, either alone or in community with others and in public or private, to manifest their religion or belief in teaching, practice, worship and observance.

Article 19.

Everyone has the right to freedom of opinion and expression; this right includes freedom to hold opinions without interference and to seek, receive and impart information and ideas through any media and regardless of frontiers.

Article 20.

(1) Everyone has the right to freedom of peaceful assembly and association.

(2) No one may be compelled to belong to an association.

Article 21.

(1) Everyone has the right to take part in the government of their country, directly or through freely chosen representatives.

(2) Everyone has the right of equal access to public service in their country.

(3) The will of the people shall be the basis of the authority of government, this will shall be expressed in periodic and genuine elections which shall be by universal and equal suffrage and shall be held by secret vote or by equivalent free voting procedures.

Article 22.

Everyone, as a member of society, has the right to social security and is entitled to realization, through national effort and international co-operation and in accordance with the organization and resources of each State, of the economic, social and cultural **rights** indispensable for their dignity and the free development of their personality.

Article 23.

(1) Everyone has the right to work, to free choice of employment, to just and favorable conditions of work and to protection against unemployment.

(2) Everyone, without any discrimination, has the right to equal pay for equal work.

(3) Everyone who works has the right to just and favorable remuneration ensuring for themselves and their family an existence worthy of **human** dignity, and supplemented, if necessary, by other means of social protection.

(4) Everyone has the right to form and to join trade unions for the protection of their interests.

Article 24.

Everyone has the right to rest and leisure, including reasonable limitation of working hours and periodic holidays with pay.

Article 25.

(1) All persons have the right to a standard of living adequate for the health and well-being of themselves and of their family, including food, clothing, housing and medical care and necessary social services, and the right to security in the event of unemployment, sickness, disability, widowhood, old age or other lack of livelihood in circumstances beyond their control.

(2) Motherhood and childhood are entitled to special care and assistance. All children, whether born in or out of wedlock, shall enjoy the same social protections.

Article 26.

(1) Everyone has the right to education. Education shall be free, at least in the elementary and fundamental stages. Elementary education shall be compulsory. Technical and professional education shall be made generally available and higher education shall be equally accessible to all on the basis of merit.

(2) Education shall be directed to the full development of the **human** personality and to the strengthening of respect for **human rights** and fundamental freedoms. It shall promote understanding, tolerance and friendship among all nations, racial or religious groups, and shall further the activities of the United Nations for the maintenance of peace.

(3) Parents have a prior right to choose the kind of education that shall be given to their children.

Article 27.

(1) Everyone has the right freely to participate in the cultural life of the community, to enjoy the arts and to share in scientific advancement and its benefits.

(2) All persons have the right to the protection of the moral and material interests resulting from any scientific, literary or artistic production of which they are the author.

Article 28.

Everyone is entitled to a social and international order in which the **rights** and freedoms set forth in this **Declaration** can be fully realized.

Article 29.

(1) All persons have duties to the community in which alone the free and full development of their personality is possible.

(2) In the exercise of their **rights** and freedoms, everyone shall be subject only to such limitations as are determined by law solely for the purpose of securing due recognition and respect for the **rights** and freedoms of others and of meeting the just requirements of morality, public order and the general welfare in a democratic society.

(3) These **rights** and freedoms may in no case be exercised contrary to the purposes and principles of the United Nations.

Article 30.

Nothing in this **Declaration** may be interpreted as implying for any State, group or person any right to engage in any activity or to perform any act aimed at the destruction of any of the **rights** and freedoms set forth herein.

THE EARTH CHARTER

PREAMBLE

We stand at a critical moment in Earth's history, a time when humanity must choose its future. As the world becomes increasingly interdependent and fragile, the future at once holds great peril and great promise. To move forward we must recognize that in the midst of a magnificent diversity of cultures and life forms we are one human family and one Earth community with a common destiny. We must join together to bring forth a sustainable global society founded on respect for nature, universal human rights, economic justice, and a culture of peace. Towards this end, it is imperative that we, the peoples of Earth, declare our responsibility to one another, to the greater community of life, and to future generations.

Earth, Our Home

Humanity is part of a vast evolving universe. Earth, our home, is alive with a unique community of life. The forces of nature make existence a demanding and uncertain adventure, but Earth has provided the conditions essential to life's evolution. The resilience of the community of life and the well-being of humanity depend upon preserving a healthy biosphere with all its ecological systems, a rich variety of plants and animals, fertile soils, pure waters, and clean air. The global environment with its finite resources is a common concern of all peoples. The protection of Earth's vitality, diversity, and beauty is a sacred trust.

The Global Situation

The dominant patterns of production and consumption are causing environmental devastation, the depletion of resources, and a massive extinction of species. Communities are being undermined. The benefits of development are not shared equitably and the gap between rich and poor is widening. Injustice, poverty, ignorance, and violent conflict are widespread and the cause of great suffering. An unprecedented rise in human population has overburdened ecological and social systems. The foundations of global security are threatened. These trends are perilous - but not inevitable.

The Challenges Ahead

The choice is ours; form a global partnership to care for Earth and one another or risk the destruction of ourselves and the diversity of life. Fundamental changes are needed in our values, institutions, and ways of living. We must realize that when basic needs have been met, human development is primarily about being more, not having more. We have the knowledge and technology to provide for all and to reduce our impacts on the environment. The emergence of a global civil society is creating new opportunities to build a democratic and humane world. Our environmental, economic, political, social and spiritual challenges are interconnected, and together we can forge inclusive solutions.

Universal Responsibility

To realize these aspirations, we must decide to live with a sense of universal responsibility, identifying ourselves with the whole Earth community as well as our local communities. We are at once citizens of different nations and of one world in which the local and global are linked. Everyone shares responsibility for the present and future well-being of the human family and the larger living world. The spirit of human solidarity and kinship with all life is strengthened when we live with reverence for the mystery of being, gratitude for the gift of life and humility regarding the human place in nature.

We urgently need a shared vision of basic values to provide an ethical foundation for the emerging world community. Therefore, together in hope we affirm the following interdependent principles for a sustainable way of life as a common standard by which the conduct of all individuals, organizations, businesses, governments, and transnational institutions is to be guided and assessed.

PRINCIPLES

I. RESPECT AND CARE FOR THE COMMUNITY OF LIFE

1. Respect Earth and life in all its diversity.

    a.   Recognize that all beings are interdependent and every form of life has value regardless of its worth to human beings.

b. Affirm faith in the inherent dignity of all human beings and in the intellectual, artistic, ethical, and spiritual potential of humanity.

2. Care for the community of life with understanding, compassion, and love.

a. Accept that with the right to own, manage, and use natural resources comes the duty to prevent environmental harm and to protect the rights of people.
b. Affirm that with increased freedom, knowledge, and power comes increased responsibility to promote the common good.

3. Build democratic societies that are just, participatory, sustainable, and peaceful.

a. Ensure that communities at all levels guarantee human rights and fundamental freedom and provide everyone an opportunity to realize his or her full potential.
b. Promote social and economic justice, enabling all to achieve a secure and meaningful livelihood that is ecologically responsible.

4. Secure Earth's bounty and beauty for present and future generations

a. Recognize that the freedom of action of each generation is qualified by the needs of future generations.
b. Transmit to future generations values, traditions, and institutions that support the long-term flourishing of Earth's human and ecological communities.

In order to fulfill these four broad commitments, it is necessary to:

II ECOLOGICAL INTEGRITY

5. Protect and restore the integrity of Earth's ecological systems, with special concern for biological diversity and the natural processes that sustain life.

    a.   Adopt at all levels sustainable development plans and regulations that make environmental conservation and rehabilitation integral to all development initiatives.

    b.   Establish and safeguard viable nature and biosphere reserves, including wild lands and marine areas, to protect Earth's life support systems, maintain biodiversity, and preserve our natural heritage.

    c.   Promote the recovery of endangered species and ecosystems.

    d.   Control and eradicate non-native or genetically modified organisms harmful to native species and the environment, and prevent introduction of such harmful organisms.

    e.   Manage the use of renewable resources such as water, soil, forest products, and marine life in ways that do not exceed rates of regeneration and that protect the health of ecosystems.

    f.   Manage the extraction and use of non-renewable resources such as minerals and fossil fuels in ways that minimize depletion and cause no serious environmental damage.

6. Prevent harm as the best method of environmental protection and, when knowledge is limited, apply a precautionary approach.

    a.   Take action to avoid the possibility of serious or irreversible environmental harm even when scientific knowledge is incomplete or inconclusive.

    b.   Place the burden of proof on those who argue that a proposed activity will not cause significant harm, and make the responsible parties liable for environmental harm.

    c.   Ensure that decision making addresses the cumulative, long-term, indirect, long distance, and global consequences of human activities.

    d.   Prevent pollution of any part of the environment and allow no build-up of radioactive, toxic, or other hazardous substances.

    e.   Avoid military activities damaging to the environment.

7. Adopt patterns of production, consumption, and reproduction that safeguard Earth's regenerative capacities, human rights, and community well-being.

    a.   Reduce, reuse, and recycle the materials used in production and consumption systems, and ensure that residual waste can be assimilated by ecological systems.

    b.   Act with restraint and efficiency when using energy, and rely increasingly on renewable energy sources such as solar and wind.

    c.   Promote the development, adoption, and equitable transfer of environmentally sound technologies.

    d.   Internalize the full environmental and social costs of goods and services in the selling price, and enable consumers to identify products that meet the highest social and environmental standards.

    e.   Ensure universal access to health care that fosters reproductive health and responsible reproduction.

    f.   Adopt lifestyles that emphasize the quality of life and material sufficiency in a finite world.

8. Advance the study of ecological sustainability and promote the open exchange and wide application of the knowledge acquired.

    a.   Support international scientific and technical cooperation on sustainability, with special attention to the needs of developing nations.

    b.   Recognize and preserve the traditional knowledge and spiritual wisdom in all cultures that contribute to environmental protection and human well-being.

    c.   Ensure that information of vital importance to human health and environmental protection, including genetic information, remains available in the public domain.

III. SOCIAL AND ECONOMIC JUSTICE

9. Eradicate poverty as an ethical, social, and environmental imperative.

  a. Guarantee the right to potable water, clean air, food security, uncontaminated soil, shelter, and safe sanitation, allocating the national and international resources required.
  b. Empower every human being with the education and resources to secure a sustainable livelihood, and provide social security and safety nets for those who are unable to support themselves.
  c. Recognize the ignored, protect the vulnerable, serve those who suffer, and enable them to develop their capacities and to pursue their aspirations.

10. Ensure that economic activities and institutions at all levels promote human development in an equitable and sustainable manner.

  a Promote the equitable distribution of wealth within nations and among nations.
  b. Enhance the intellectual, financial, technical, and social resources of developing nations, and relieve them of onerous international debt.
  c. Ensure that all trade supports sustainable resource use, environmental protection, and progressive labor standards.
  d. Require multinational corporations and international financial organizations to act transparently in the public good, and hold them accountable for the consequences of their activities.

11. Affirm gender equality and equity as prerequisites to sustainable development and ensure universal access to education, health care, and economic opportunity.

  a. Secure the human rights of women and girls and end all violence against them.
  b. Promote the active participation of women in all aspects of economic, political, civil, social, and cultural life as full and equal partners, decision makers, leaders, and beneficiaries.

c.   Strengthen families and ensure the safety and loving nurture of all family members.

12. Uphold the right of all, without discrimination, to a natural and social environment supportive of human dignity, bodily health, and spiritual well-being, with special attention to the rights of indigenous peoples and minorities.

a.   Eliminate discrimination in all its forms, such as that based on race, color, sex, sexual orientation, religion, language, and national, ethnic or social origin.
b.   Affirm the right of indigenous peoples to their spirituality, knowledge, lands and resources and to their related practice of sustainable livelihoods.
c.   Honor and support the young people of our communities, enabling them to fulfill their essential role in creating sustainable societies.
d.   Protect and restore outstanding places of cultural and spiritual significance.

IV.   DEMOCRACY, NONVIOLENCE, AND PEACE

13. Strengthen democratic institutions at all levels, and provide transparency and accountability in governance, inclusive participation in decision making, and access to justice.

a.   Uphold the right of everyone to receive clear and timely information on environmental matters and all development plans and activities which are likely to affect them or in which they have an interest.
b.   Support local, regional and global civil society, and promote the meaningful participation of all interested individuals and organizations in decision making.
c.   Protect the rights to freedom of opinion, expression, peaceful assembly, association, and dissent.
d.   Institute effective and efficient access to administrative and independent judicial procedures, including remedies and redress for environmental harm and the threat of such harm.
e.   Eliminate corruption in all public and private institutions.
f.   Strengthen local communities, enabling them to care for their environments, and assign environmental responsibilities to

the levels of government where they can be carried out most effectively.

14. Integrate into formal education and life-long learning the knowledge, values, and skills needed for a sustainable way of life.

a. Provide all, especially children and youth, with educational opportunities that empower them to contribute actively to sustainable development.

b. Promote the contribution of the arts and humanities as well as the sciences in sustainability education.

c. Enhance the role of the mass media in raising awareness of ecological and social challenges.

d. Recognize the importance of moral and spiritual education for sustainable living.

15. Treat all living beings with respect and consideration.

a. Prevent cruelty to animals kept in human societies and protect them from suffering.

b. Protect wild animals from methods of hunting, trapping, and fishing that cause extreme, prolonged, or avoidable suffering.

c. Avoid or eliminate to the full extent possible the taking or destruction of non-targeted species.

16. Promote a culture of tolerance, nonviolence, and peace.

a. Encourage and support mutual understanding, solidarity, and cooperation among all peoples and within and among nations.

b. Implement comprehensive strategies to prevent violent conflict and use collaborative problem solving to manage and resolve environmental conflicts and other disputes.

c. Demilitarize national security systems to the level of a non-provocative defense posture, and convert military resources to peaceful purposes, including ecological restoration.

d. Demilitarize national security systems to the level of a non-provocative defense posture, and convert military resources to peaceful purposes, including ecological restoration.

e. Ensure that the use of orbital and outer space support environmental protection and peace.

f. Recognize that peace is the wholeness created by right relationships with oneself, other persons, other cultures, other life, Earth, and the larger whole of which all are a part.

THE WAY FORWARD

As never before in history, common destiny beckons us to seek a new beginning. Such renewal is the promise of these Earth Charter principles. To fulfill this promise, we must commit ourselves to adopt and promote the values and objectives of the Charter.

This requires a change of mind and heart. It requires a new sense of global interdependence and universal responsibility. We must imaginatively develop and apply the vision of a sustainable way of life locally, nationally, regionally, and globally. Our culture diversity is a precious heritage and different cultures will find their own distinctive ways to realize the vision. We must deepen and expand the global dialogue that generated the Earth Charter, for we have much to learn from the ongoing collaborative search for truth and wisdom.

Life often involves tensions between important values. This can mean difficult choices. However, we must find ways to harmonize diversity with unity, the exercise of freedom with the common good, short-term objectives with long-term goals. Every individual, family, organization, and community has a vital role to play. The arts, sciences, religions, educational institutions, media, businesses, nongovernmental organizations, and governments are all called to offer creative leadership. The partnership of government, civil society, and business is essential for effective governance.

In order to build a sustainable global community, the nations of the world must renew their commitment to the United Nations, fulfill their obligations under existing international agreements, and support the implementation of Earth Charter principles with an international legally binding instrument on environment and development.

Let ours be a time remembered for the awakening of a new reverence for life, the firm resolve to achieve sustainability, the quickening of the struggle for justice and peace, and the joyful celebration of life.

THE SCIENCE OF BEING ALTOGETHER

The past, present, and future are all contained in the Life Force.
Morihei Ueshiba, *The Art of Peace*

Artists are the unacknowledged legislators of the world.
Percy Byssche Shelley

What is right and good in our activities is what is in harmony with the workings of the universe, or the 'Life Force'. We know what is universal harmony because we belong to it by the nature of our being. In other words, each one of us is a participant in the Life Force from the moment of our birth to our death and beyond. Each one of us *knows* harmony. It is the peace that passes all understanding. It is our birthright. It is where we belong. Some call the workings of the universe 'god' or 'allah' or "jehovah", "hari", "brahman". Submission to the will of God in this sense is synonymous with the dedicated desire to act in harmony with the workings of the universe, to act in a right and good way as intuited by our being and implemented by our will. What does this truly mean?

For the Australian aborigines and the Native Americans there was little in the way of overlay such as cities, governments, conflicting world views, or even buildings, railroad tracks or roads on the land to stand between them and the first hand experience of the workings of the universe. These workings include what we call "natural", such as the rising of the sun and moon, the participation of the stars and clouds in the environment and the remorselessness of the other elements as rainfall and temperature. In this simplicity of life there is great clarity in the understanding that mortals are integral participants in the Life Force along with all living and inanimate things. Humans understand themselves to be directly related to the other animals and respect them for their beingness with the same *reverence* that they respect themselves and other mortals.

In the science of being altogether, some constructs of civilization are in harmony and others are not, but only a study of the science itself reveals to us which is which. Generally speaking, creativity enhances harmony and wilful destruction of life and property harms it. No killing along the food chain is wilful, as the creativity of sustenance of others prevails. The greatest violation (and hence hubris) in the science of being altogether is the harming or killing of human beings by each other AND the wilful destruction of the environment. The greatest tragedy occurs when the destruction is not recognized as a *scientific* violation. In the entire length of this civilization, humans are comparatively peaceful within their own group or circle whether small or large as a clan, a region or a country. Yet, the other(s), or those outside of this grouping, have been viewed with distrust, hostility and often murderous ferociousness, through the ages, without abatement, to the present moment.

Because of their intelligence, human beings have been able to override the workings of the universe in the flight toward exploring the ever-expanding reaches of their minds. The lessons learned and the hard-earned understanding of the ways of the workings of the universe, inherent in the cultures of the Australian aborigines in an arid land and the Native Americans in a fertile one, have been misinterpreted or deliberately ignored and degraded to such an extent that to re-understand them appears to be almost impossible. Yet without such understanding, no human endeavor however magnificent, will survive its remorseless inevitability. Scattered over the continent of Australia are the remains of structures that people thought could defy the elements. In America the healing components of a marvelously fertile land are often ignored as excursions into the great wilderness develop artificial methods of health-keeping and wealth display that are *scientifically untenable* and as such will pass into nothingness.

Thanks to the work of great scientists in the twentieth century, the universe of the twenty-first century may now be viewed as a work of art, in that the thread or string uniting all of the cosmos into a harmonious whole is recognized in the same way as the divinity in a work of art. The difference is that a work of art may be viewed in its totality. In terms of activity of human beings, the art form that aspires to represent the universe in microcosm is *tragedy**, and the gift of *tragedy* has been given to the world from days prior to the birth of Christ and up to the present day. In the great tragedies, human frailty and wilful violation of the known workings of the universe as speciel killing, are shown in the nakedness of their hubris and chastened or destroyed by the remorseless drive of the Life Force. Harmony prevails in the end. Whenever disruptions occur, the life force, or the moving unifier or even, with tolerance for the inadequacies of language, the will of God, ultimately prevails. In the sheer strength of this prevailing, we can understand on the pulses, the nature of our own participation in the Science of Being Altogether, our place in the TIMELINE of our civilization and our role today.

The patterns of behavior described at the birth of this civilization from the building of the pyramids, through the writings of the great prophets still remain unchanged at the present time. The propensity to divide humanity into two parts or two sides, as friend and enemy, good and bad, continues to wreak destruction on the planet in the same way that it did 2000 years ago at the dawn of this so-called western civilization which is now merging into the first global civilization we believe the planet has ever known.

* See Other Definitions

Free spirited artists have never been silent
but almost always ignored.
However,
never in the history of human kind has there been such a clear awareness
that just as art imitates life
so may life now imitate art in a practical sense,
to alter patterns of behavior that have been continually repeated
for over twenty centuries
and have never resulted in the peace that passes all understanding ,
which is advocated by all of the major religions
as well as most of the governments
that have ever existed.
From the still point of our own self, guided by our own sense of
inner harmony,
or conscience,
or our own immediate relationship with our creator, our divinity,
our own unique participation in the Life Force,
we move outward,
into our immediate family, our extended family, our society, our nation,
our culture,
our commercial enterprises, our global civilization, our participation in
the eco-system,
the cosmos,
understanding
our belongingness to the collective consciousness of our human past and
our human future,
and we move always guided,
incorporating all of these things into the requirements of our conscience,
our birthright,
that can never be taken from us,
and can only be violated by each one of us, our self, alone.

M.A.M.

RELIGIONS AND SCRIPTURES

Some Notes on the World's

Living Religions , Spiritual Movements

and Related Individuals,

including their Sacred Writings*

At the Dawn

Of the New Millennium

* Materials are adapted from *WORLD SCRIPTURES*, edited and compiled by Rev. Leland P. Stewart, in cooperation with leaders of the various faith communities.

RELIGION - WHAT IS IT?

"Religion" comes from the Latin word "religio", which means to bind, to unify.

Religions traditionally have been initiated by single enlightened beings who offer an example of a life of vision and faith. They are accompanied by teachings around which a community is formed over time. Generally speaking, religions are based on the life and teachings of these individuals extended over time, both during the lifetime of the founders and beyond.

Denominations are quite different from religions. They are divisions within a religion. Presbyteranism and Methodism are denominations within Protestant Christianity. Some Christian denominations consider themselves independent; thus they do not relate to any overall group of churches.

Architectural buildings such as churches are often used interchangeably with the religion which is practiced there. However, many of the more recent spiritual movements may have their place located in a storefront or even a house. Many practitioners of established religions also may share a large church building with other denominations or religions.

In addition to the historic religions, there are modern spiritual movements. Whereas, in the past, longevity has had a significant part in determining what is considered to be a "religion", there are also modern movements which consider themselves to be religions. Baha'i is a good example. It began during the nineteenth century and has many powerful teachings.

The native traditions of indigenous peoples, such as those of the North American Indians and the Australian aborigines, have a long oral history and development of community. In this sense they are considered to be ancient religions whose teachings are only recently becoming recognized as significant as the world moves towards global harmony.

An article on money is included. Money, though devoid of inherent spirituality, has qualities of adherence which may surpass, and have surpassed , those qualities of unifying which are intrinsic in the great religions and spiritual movements.

TAOISM

According to Chinese tradition, Taoism (pronounced Dowism) originated with a man named Lao Tze, said to have been born about 604 B.C.E. He was probably an older contemporary of Confucius. Scholars know nothing for certain about him and often wonder if he actually lived. Legend indicates that he was saddened by his people's disinclination to cultivate the natural goodness he taught, so he climbed on a water buffalo and rode westward toward what is now Tibet. As he was about to leave China, a gatekeeper tried to persuade him to turn back. Lao Tze refused, but when he was asked to write down the teachings he was advocating, he returned in about three days with the *Tao-Te-Ching*. This slim volume is even today considered to be the basic text of Taoist thought.

Chuang Tze belongs to the third and fourth centuries B.C.E., being the major disciple of Lao Tze. He has had a corresponding influence on Taoism as St. Paul had upon Christianity. He wrote *The Works of Chuang Tze*. It is these two documents which have shaped the religion called Taoism ever since. Because of the abstractness of these two texts, the religion of Taoism has not lived up to the loftiness of the teachings, but rather has become overladen with rather superstitious rituals and ceremonies. Nevertheless, there is much wisdom in these teachings which is applicable to the world as a whole.

ABOUT THE SCRIPTURES

Tao-Te-Ching

Tao-Te-Ching contains the very profound and beautiful, yet applicable religious, political and philosophical teachings of China's Lao Tze. Lofty mysticism and sound ethics are both to be found in this document. The ideas of faith and love are here expressed in fundamentally the same way as Jesus stated them more than five centuries later. Taoism was chosen to be first in the *World Scriptures* because of its beautiful expression of how life and the world began.

> There is a thing, inherent and natural,
> Which existed before the universe.
> Motionless and fathomless,
> It stands alone and never changes;
> It pervades everywhere and never becomes exhausted.
> It may be regarded as the Essence of the universe..

The Works of Chuang Tze

Chuang Tze elaborated the principles of the *Tao-Te-Ching* in a prosaic but profound manner. "The Way of the Good Carpenter", for example, gives an illustration of the importance of meditation prior to creating a work of art, in this case a stand for musical instruments. The content of these documents contain a very pronounced mystical element, but ethics also are developed in the teachings

L.P.S.

CONFUCIANISM

Confucius was born in 551 B.C.E. in the province of Shantung, China. He married at nineteen and devoted his life to teaching and government. For a certain period he was actually in political life, but afterward he chose to teach about good government and the development of culture in China. His emphasis is shown to have been predominantly ethical, but it holds definite value for the world today.

Taoism and Confucianism together give a good illustration of the two basic elements of religion: mysticism and ethics. These two do not necessarily contradict one another, but in the past they often have done so. The proper balance is what is difficult to attain.

ABOUT THE SCRIPTURES

Analects of Confucius

The *Analects of Confucius* consists of nineteen books of Confucius's teachings in a rather loose and informal style. Most of the teachings are given in dialogue form with Confucius, such as questions and comments from other people. They include ethics and desired forms of behavior. Confucius accepted the notion of spiritual reality but did not deal with it in his teachings. Instead, he emphasized virtue and mutual respect, as well as other ways to live that are based upon the ethics accepted by Chinese society at its best.

The Great Learning

The *Great Learning* is a brief teaching "to illustrate illustrious virtue; to renovate the people, and to rest in the highest excellence". The cultivation of the person is the root of everything.

The Doctrine of the Steadfast Mean

Considered one of the central teachings of Confucianism, the steadfast mean is that "being without inclination to any excess is called balance; admitting of no change is called steadfastness". "...superior people cultivate a friendly harmony, without being weak. How firm they are in their energy! They stand erect in the middle without inclining to either side".

The Works of Mencius

Mencius was the most outstanding disciple of Confucius, although the followers of Confucius were quite numerous. The writings are about the teachings of the Master but were written by his disciples. Once again they deal with ethical questions.

The Book of Filial Piety

Filial piety has to do with respect for our elders, including our parents, teachers, and even governmental officials. "The duty of children to their parents is the fountain whence all other virtues spring, and also the starting point from which we ought to begin our education". It is acceptable to resist our parents or other elders if what they wish us to do is wrong, but in general we should learn from them and follow them whenever possible.

L.P.S.

HINDUISM

Hinduism is the name used to designate the religious creeds and practices of a great number of Hindus, who form the body of the oldest major religion alive today.

ABOUT THE SCRIPTURES

Upanishads

In the eighth and/or seventh centuries B.C.E., a group of Hindu thinkers wrote down various speculations regarding their previously developed rituals, in a number of separate *Upanishads*. The *Upanishads* are the ending portion of each of the four Vedas. They are the wisdom portion and equal the end, or goal, as well as the final portion of the *Vedas*. The ritual portion contains ancient rituals, prayers, and hymns.

Bhagavad-Gita

The gigantic epic poem, *The Mahabharata*, which was written over a period of several hundred years (perhaps 500 B.C.E. to 250 B.C.E.), contained the *Bhagavad-Gita*, which has become the most prominent scripture for Hindus as a whole. The *Gita* portrays the battle of life, and is presented here as a discourse between the great teacher Krishna and his disciple Arjuna. This document brings out very clearly the variety of belief and practice which is possible in searching for the Eternal.

Sankaracharya's Atma Bodha

This short writing gives some idea of how it is possible to discover the meaning of the soul.

The Yoga Sutras of Patanjali

The Hindus, in their yoga systems, have the most highly developed science of human discipline ever known to humans. Western psychology and medicine are just today discovering how much these people learned by means of yoga about which technological science is still ignorant. Patanjali's system, with very little interpretation, yields a method of Self-realization that in general principle is universally applicable.

L.P.S.

BUDDHISM

Buddhism as a religion developed from the teachings of Gautama, a prince of the Sakya Clan in India, who lived from 560 to 480 B.C.E. At the age of twenty-nine Gautama left his palace, and his wife and son, in order to search for a way by which humanity could overcome the suffering and misery of life. After many years of asceticism and renunciation, he finally reached a satisfactory set of conclusions, which he proceeded to teach to a group of disciples for a period of about forty years. As a result of his enlightenment he was called *the Buddha*, meaning *the enlightened one.*

ABOUT THE SCRIPTURES

The Life and Teachings of the Buddha

The first two sections of the Buddhist Scriptures are the life and the teachings of the Buddha. The life is developed in some detail as to what stages *the enlightened one* went through in order to arrive at his decision to leave the palace and his early life in order to seek enlightenment and then to carry out his mission. The teachings are also extensive and include the Eightfold Path, the Nine Avoidances, and the Setting Up of Mindfulness.

The Dhammapada

This document is a series of proverbs on a number of religious and philosophical subjects.

The Diamond Sutra

This Sutra, is a talk supposedly given by the Buddha in answer to the fundamental questions of one of his disciples, called Sabhiti.

Asvaghosha's Discourse on the Awakening of Faith

Asvaghosha attempts here to systematize two aspects of the soul. Such an undertaking is dangerous if one does not always keep in mind that the whole is greater than the summation of its parts: ultimately it is impossible to subdivide religious ideas and retain their complete essence. At the same time, this attempt can be very fruitful if we remember that we must reassemble the parts once we have examined them.

The Tibetan Doctrine

The Tibetan Doctrine is a collection of poetic proverbs from a number of Tibetan writings.

A Manual of Zen Buddhism

Zen Buddhism perhaps owes as much credit to Taoism as to Buddhism. Whether or not this is true, it does represent an integration of the thought of these two religions. It developed to some extent in China as Chan Buddhism, but it has its most outstanding development in Japan.

L.P.S.

JAINISM

Originating on the Indian sub-continent, Jainism is one of the oldest religions of its homeland and indeed of the world. Having pre-historic origins, before 3000 B.C.E., and before the propagation of Indo-Aryan culture, it had numerous exponents from that time until the coming of Lord Mahavira (599 - 527 B.C.E.), who is considered the founder of the religion These teachings continue today in India and around the world. Jain religion is unique in that, during its existence of over 5000 years, it has never compromised on the concept of non-violence (ahimsa) either in principle or practice. Vegetarianism is a way of life for Jains, taking its origin in the concept of compassion for all living beings, *Jiva Daya*.

ABOUT THE SCRIPTURES

The scriptures of Jainism center around a code of conduct which is made up of the following five vows, and all of their logical conclusions: Ahimsa (harmlessness), Satya (truthfulness), Asteya (non-stealing), Aparigraha (non-possessiveness), and Brahmacharya (chastity).

As indicated in its scriptures, this system of religion, thought, and living stands for the highest and noblest human values, and offers a path of eternal peace and happiness.

Adapted from *Sourcebook of the World's Religions*, Ed. Joel Beversluis.

L.P.S.

ZOROASTRIANISM

Iran, at the time of Zarathustra's birth, was a land where many pagan gods and goddesses were being worshiped through ignorance and fear. The Prophet Zarathustra in his sublime hymns, the Gathas, revealed to humanity that there was the One, Supreme, All-knowing, Eternal God of the good creations, the Lord of Wisdom, who was wholly Wise, Good and Just. Ahura Mazda, he taught, was a friend to all and never to be feared by human beings, who in turn should worship him. He proclaimed the two primordial spirits were locked in open conflict. These were Spenta Mainyu, the Holy Spirit of Ahura Mazda and His diabolical adversary, Anghra Mainyu, the Hostile Spirit.

ABOUT THE SCRIPTURES

The Zoroastrian Doctrine

According to the Zoroastrian texts, Ahura Mazda (Pahlavi Ohrmazd), through his Omniscience knew of His own Goodness and His infinite Self, as well as He was aware of the hostile spirit's limited strength and finite existence. In order to destroy His adversary, Ahura Mazda created an immaculate material world of the seven creations to trap the hostile spirit. Ahura Mazda knew that the material world would bring within it disorder, falsehood, wickedness, sorrow, cruelty, disease, suffering and death. Human beings, Ahura Mazda's finest creations, are believed to be the central figures in this cosmic struggle. The prophet declared that it is during this period of conflict that humanity, through free will, should choose to fight and vanquish the hostile spirit using the ethical paradigm of Goodness, the Good Mind, Truth, Power, Devotion, Perfection and Immortality. These seven qualities collectively came to be known as the Amesha Spentas (Bounteous Immortals). It is our responsibility to imbibe the virtues of these divinities in order to know how to generate the right thoughts, words and actions. Zarathustra recognized that the use of these principles of righteous living would enable us to bring about the eventual annihilation of evil in this world.

Humanity

Humanity's unique spiritual quest, according to Zoroastrianism, is linked to the preservation and promotion of the Wise Lord's seven creations - namely, the sky, waters, earth, plants, cattle, humans and fire. The last creation, fire, is a potent reality in Zarathustra's revelation, as the prophet saw fire to be the physical representation of Asha (Order/ Truth/Righteousness), and as a source of light, warmth and life for all people. The religious rituals (the performance of which is an important Zorastrian duty) are solemnized in the presence of fire, the life-energy which permeates and makes dynamic the Wise Lord's other six creations.

Living a Zoroastrian Life

Zarathustra taught that since this world created by Ahura Mazda is essentially good, we should live well and enjoy its bountiful gifts though always in moderation, as the states of excess and deficiency in Zoroastrianism are deemed to be the workings of the hostile spirits. Humanity, in Zoroastrianism, is encouraged to lead a good life and prosperous life and hence monasticism, celibacy, fasting and the mortification of the body are anathema to the faith; such practices are seen to weaken us and thereby lessen our power to fight evil. The Prophet saw pessimism and despair as sins; in fact, as yielding to evil. In his teachings, we are encouraged to lead an active, industrious, honest and above all, a happy and charitable life.

The After-Life Doctrine

Upon physical death (which is seen as the temporary triumph of evil) the soul will be judged at the Bridge of the Separator, where the soul, it is believed, will receive its reward or punishment, depending upon the life which it has led in this world, based upon the balance of its thoughts, words and deeds. If found righteous, the soul will ascend to the abode of joy and light, while if wicked, it will descend into the depths of darkness and gloom. The after state, however, is a temporary one, as there is no eternal damnation in Zorastrianism. There is a promise, then, of a series of saviors (the Saoshyants) who will appear in the world and complete the triumph of good over evil. Evil will be rendered ineffective and Ahura Mazda, the Infinite One, will finally become truly Omnipotent in Endless Light. There will then take place a general last judgment of all the souls awaiting redemption, followed by resurrection of the physical body, which will once again meet its spiritual counterpart, the soul. Time as we know it will cease to exist, and the seven creations of Ahura Mazda will be gathered together in eternal blessedness in the Kingdom of Mazda, where everything, it is believed, will remain in a perfect state of joy and immortality.

The History

For over a thousand years, from about 625 B.C.E. to 549 B.C.E., the religion taught by Zarathustra flourished as the state religion of three mighty Iranian empires, that of the Achaemenian, the Parthians and Sasanians. Among the many subjects of the Achaemenian Empire were the Jews, who adopted some of the Prophet's main teachings and transmitted them in due course to Christianity and later to Islam.

The Parsi Arrival

In the seventh century C.E, the Arabs conquered Iran. Many of them settled there and gradually imposed their own religion of Islam. In the early first century, a small group of Zoroastrians seeking freedom of worship and economic redress, left Iran and sailed towards the warm shores of Western India. They eventually arrived along the Gujarat coastline in 936 C.E. at a place they named Sanjan, some 180 kilometers north of Bombay. There they flourished and came to be known as the Parsis (Persians). Over the millennium, a small band of faithful Zoroastrians have continued to live in Iran and have tried to preserve their culture and religious traditions as best they could.

Mobed Zarir Bhandara,
Zoroastrian High Priest

JUDAISM

Judaism has a history of its own that goes back into the dim past in the Middle East. One great theme dominates the course of Jewish religion. This is the theme that a single, righteous God is at work in the social and natural order. Being socially minded, the Jews were historically aware, not in a casual or intermittent way but steadily. Moses emerged as the central figure, yet there are other prophets who also gave real strength and direction to the development of Judaism. The Hebrew Scriptures serve as a way to put forth their teachings and also to record their history as completely as they could.

L.P.S.

ABOUT THE SCRIPTURES

The Hebrew Scriptures comprise thirty-nine books in three divisions: Torah, Prophets, and Writings. These selections are from the Jewish Publication Society (1950 edition). They were adapted in modern English by the Editor. (Thirty-nine is a popular, non-canonical enumeration.)

The Torah contains the five books of Moses, the Prophets twenty-one books, and the Writings thirteen books. Altogether, their authorship spans eight centuries, from the thirteenth century B.C. E. to the fifth century B.C.E.

The *Mishnah* is a compendium of Jewish law and was formulated by the sages in ancient Judea and edited by Judah the Prince 220 B.C.E. in six divisions, which became the basis for Talmudic discussion. The selection here is from Mishnah Abot, The Ethics of the Fathers, a collection of wise sayings in six chapters but without Talmudic commentary remaining open to public discourse for generations to come.

The *Midrash Aggadah*, as distinguished from Midrash Halachah (Law) is Rabbinic interpretation of homiletical, metaphorical, and philosophical nature. Since there are Aggadic elements in the Torah, Midrash is indigenous to Jewish learning from ancient to modern times if and only if new interpretation is grounded in Scripture. Representing the Midrashic mode of Torah commentary is the *Treatise on Peace* (c. 6th-7th century B.C.E.). This minor treatise on "Proper Decorum" can be found in the *Vilna Talmud* as "Derekh Eretz Zuta".

The criteria for selection were as follows: (I) limits of space, (II) covenantal meaning; (III) literary form, and (IV) universal significance.

The historic experience of Israel reflects a growing self-awareness of its role as a People of God. The classical sources are defining documents of its character and destiny, testifying to this spiritual evolution as a continuous struggle for meaning and transcendence.

Wisdom dictates a method of interpretation which avoids literal acceptance of the sacred text if it goes contrary to human experience, natural law and the principles of reason. The allegorical method is that Midrashic tool for overcoming contradictions between language and reality, provided its own constructions are not taken literally. The confusion of epistemological categories of thought is the major source of superstition and falsehood. The human imagination, while in itself a miracle of creative wonder, can become a dangerous force if unbridled and undisciplined, especially in the domain of metaphysics and theology.

Scripture as sacred text inspired by Divine Influence, is a record of the human response to the mystery of the Divine Reality as expressed in human language. The Word of God is the word about God or addressed to and from God in a cultural context, and each generation has its own interpreters according to its experience and comprehension. Faith in God is pure intuition of trust, not a sanction for confusion, dogma, superstition and bigotry, nor is it a shortcut to true knowledge or a substitute for it. What then is its ultimate value?

The notion of Covenant is one of commitment to the Divine Reality as a joint presence in the life of the Spirit. Biblically, there are five covenants between God, Israel and Humanity, namely the Covenant of Creation (the Sabbath); the Covenant of Humanity (Noah): the Covenant of Land and People (Abraham): the Covenant of Revelation (Torah at Sinai); and the Covenant of Peace (Universal Redemption).

The five Covenants are essentially one in form if not in content. To unite form and content is the historical task. To be chosen is to teach that every human being is thus chosen as a beloved of God.

The seventh century (B.C.E.) Prophet Zephaniah has expressed the vision of the Covenant of Peace: (CH 3:9):

For then will I turn to the peoples
A pure language,
That they may all call upon the name of the Lord,
To serve Him with one consent.

Rabbi Benjamin Herson

CHRISTIANITY

The religion known as Christianity began through the life and death of Jesus of Nazareth, who was really unknown until his ministry began at around the age of thirty, and lasted only about three years after that time. The stories of his meeting with the rabbis in the temples at the age of twelve appear to be legends that could have happened but were not recorded at the time they occurred. Jesus was a Jew, and the teachings which came after him that formed the New Testament were built upon the Old Testament (or Hebrew Bible) that already was well established with the Jews.

It was when Jesus became the Christ, and especially when his followers proclaimed him as Lord and Savior, that most Jews chose to return to their own faith and accept that Christianity was truly a separate religion. The death and resurrection of Jesus gave it a uniqueness that has often been used to see Christianity as unique among all the religions of the world.

ABOUT THE SCRIPTURES

Old Testament

All of the Christian Scriptures are, directly or indirectly, of Hebrew origin. The Old Testament itself contains the basis of historic Judaism, particularly in the Pentateuch (or the first five books of the Bible). Christians have adopted the Old Testament as an essential part of their canon, but it is supplemented by the books of the New Testament.

Judaism traces its origin to the life and teachings of Moses around 1200 B.C.E., when he freed the Israelites from captivity in Egypt and established them in the worship of Jehovah. Their religion, which was called "the religion of Israel", eventually became so universal in its conceptions that it is still able to hold the Jews together, even though they are scattered throughout the world.

Christians see the Old Testament as essential to their faith because Jesus was a Jew, and because the roots of Christian faith come from many of the teachings and prophecies of the Hebrew Bible. Thus the same scriptural material is used by both religions though often seen in a different light.

New Testament
The Life and Teaching of Jesus of Nazareth

The New Testament in *World Scriptures* contains a unified picture of the life and teachings of Jesus, as told in the four gospels, but interpreted in a modern light. Here, while Jesus is still the great figure of the Palestinian setting, he speaks in terms meaningful in the world of today, thereby showing the result of approximately two thousand years of slow, but eventually inevitable, progress in religious thought. The term "Christ" is used to mean the enlightened individual who is thought by Christians to have fulfilled the Jewish messianic prophecies; this interpretation makes the notion approximately parallel to the term "Buddha" in the Buddhist tradition.

Probably the most prominent single idea which Jesus taught was that of the Kingdom of God. To him it was conceived as a supernatural kingdom which would come at the expected end of the natural world. Although Jesus was incorrect in his notion that the world would end "within a generation", he did have many timeless insights about the Kingdom and about the people who would be part of it. However, instead of the Kingdom's being a social utopia on earth, as many modern Christians have concluded, it is much more adequate as an inner condition of lasting happiness, which should be apparent in the parables and other teachings.

Other New Testament Writings

The books of the Christian Bible (the *New Testament*) that follow the gospels show the process of change which Christianity underwent in becoming adapted to the needs of many different kinds of people. Of course, most of the writings contain some doctrine suitable to this particular religion. Perhaps the best way to describe the essential worth which these books contribute to religion is to say that they express the perfectionist ethics of Jesus in a variety of ways which are more nearly attainable by the society than were the original ideals themselves. Relatively little mysticism is present, as compared with the Bhagavad Gita and the Tao-Te-Ching; rather, the stress is upon God's love as applied to daily living.

The first section, entitled *Paul the Apostle*, gives a general picture of the towns and cities which Paul, the man most responsible for the establishment of the Christian Church, visited while at his various locations. Also, two of Paul's co-workers are given advice as to what their own teachings should be. Whether modern scholarship finds all of these words to be Pauline or not makes no difference in this interpretation. The important item is the present worth of the teachings themselves.

Later sections are in letter form and comprise those writings of other disciples and followers of Jesus which still have outstanding value for today.

L.P.S.

ISLAM

*(*Adapted with permission from Shaikh, Munir A. *Teaching About Islam and Muslims in the Public School Classroom.* Council on Islamic Education, Fountain Valley, 1995).

The term *Islam* derives from the three-letter Arabic root s-i-m, which generates words with interrelated meanings, including "surrender", "submission", "commitment", and "peace". Commonly, Islam refers to the monotheistic religion revealed to Muhammad ibn (son of) Abdullah (pbuh) between 610 and 632 of the Common Era. The name Islam was instituted by the Qur'an, the sacred scripture revealed to Muhammad. For believers, Islam is not a new religion. Rather, it represents the last reiteration of the primordial message of God's Oneness, a theme found in earlier monotheistic religious traditions.

Although Islam can be described as a religion, it is viewed by its adherents in much broader terms. Beyond belief in specific doctrines and performance of important ritual acts, Islam is practiced as a complete and natural way of life, designed to bring God into the center of one's consciousness, and thus one's life. Essentially, by definition Islam is a world view focused on belief in the One God and commitment to His commandments. (Shaikh, 1995)

ABOUT THE SCRIPTURES
What does the term Allah mean?

The Arabic word "Allah" is a contraction of the words "al" and "ilah", and literally means "The God". Believers in Islam understand Allah to be the proper name for the Creator as found in the Qur'an. The name "Allah" is analogous to "Eloh", a Semitic term found in the divine scriptures revealed to Muhammad's predecessors Moses and Jesus (may peace be upon them all).

The use of the term "Allah" is not confined to believers in Islam alone. Allah means "God", as "Dios" and "Dieu" mean "God" in Spanish and French respectively. It is common to find many Arabic-speaking Christians and Jews also using the word Allah in reference to God. (Shaikh,. 1995).

How is God viewed in Islam?

The Qur'an, the divinely-revealed scripture of Islam, contains numerous verses describing the nature of God. The role of human beings as creations of God upon the earth and their relationship with God are also discussed extensively in the sacred text.

> Say: He is God, the One, the Eternal, Absolute. He does not beget, nor is He begotten, and there is nothing like unto Him. (I12:1-4)
> It is He who brought you forth from the wombs of your mothers when you knew nothing and He gave you hearing and sight and intelligence and affections that you may give thanks. (16:78)

> No vision can grasp Him, but His grasp is over all vision. He is above all comprehension, yet is acquainted with all things. (6:103)

Muslims believe that God has no partners or associates who share in His divinity or authority. Muslims also believe that God is transcendent and unlike His creations, and thus has no physical form. Nor is God believed to exist in (or be represented by) any material object. A number of divine attributes or names, which serve to describe God, are found in the Qur'an. Some commonly known attributes include the Most Merciful, the Most Forgiving, the Most High, the Unique, and the Everlasting, among others.

In Islam, human beings, like other creations, are seen as completely unlike God, though they may aspire to exhibit various attributes manifested by God, such as justice or mercy. Furthermore, even while God is believed to be beyond traditional human perception, the Qur'an states, "He is with you wherever you may be." (57:5). For Muslims, God's oneness heightens the awareness that all life is ultimately bound by Divine law emanating from a singular source and that life has a meaning and purpose which revolves around the consciousness of God's presence.

Moreover, belief in a singular creator compels conscientious Muslims to view all humanity as one extended family, and treat others with justice and equity. Respect for the environment and natural resources also follows from the Muslim view of God.(Shaikh, 1995).

What is the Qur'an?

The word Qur'an literally means "the reading" or "the recitation", and refers to the divinely revealed scripture given to Muhammad. Since Muhammad is considered the last prophet of God, the Qur'an is believed to be the final revelation from God to humanity.

The Qur'an is considered by Muslims to be the literal Speech of God given to Muhammad in the Arabic language. The chapters and verses of the Qur'an were revealed throughout Prophet Muhammad's mission, over a span of close to twenty-three years, from 610-632 B.C.E. Contrary to common misconception, Muhammad is not the author of the Qur'an.

Rather, he is viewed as the chosen transmitter of the revelation and the ideal implementer of principles and commandments contained therein. The personal sayings or words of Muhammad are known as hadith, which are distinct from the divine origin of the content of the Qur'an.

As verses of the Qur'an were revealed to Muhammad and subsequently repeated by him to companions and other fellow Muslims, they were written down, recited, and memorized. The Prophet also typically led the formal worship five times daily, during which he recited the revealed verses according to the procedure that he established. The verses were also recited out loud by designated Muslims in the early dawn hours and prior to the worship times and other important occasions. In short, the Qur'anic verses played an immediate and practical role in the spiritual lives of Muslims from the outset. Before he passed away, the Prophet arranged the 114 chapters into the sequence we find in the Qur'an. Scholars, both Muslim and non-Muslim, agree that the Qur'an has remained intact and unchanged to the present. (Shaikh, 1995) *Aiyub Palmer*

Do Translations of the Qur'an exist in other languages?

Translations of the Qur'an exist in many languages throughout the world, including English, Spanish, French, German, Urdu, Chinese, Malay, Vietnamese, and others. It is important to note that while translations are useful as renderings or explanations of the Qur'an, only the original Arabic text is considered to be the Qur'an itself. As a consequence, Muslims the world over, regardless of their native language, always strive to learn Arabic, so they can read and understand the Qur'an in its original form. Muslims also learn Arabic in order to recite the daily formal worship (salah) and for greeting one another with traditional expressions. However, while almost all Muslims have some basic familiarity with the Arabic language, not all Muslims speak fluent Arabic. (Shaikh, 1995)

Who was Muhammad?

History records that a person by the name of Muhammad was born into the tribe of Quraysh in the city of Makkah in 570 C.E. His father, Abdullah, fell ill and died before his birth. When Muhammad was six years old, his mother, Amina, also became ill and died. Thus at a very young age Muhammad was orphaned by the loss of both his parents.

For the next two years Muhammad was entrusted to his grandfather, Abd al-Muttalib, who passed away of old age. His uncle, Abu Talib, a well-respected member of the Quraysh tribe, took responsibility for young Muhammad. Muhammad grew up to become an honest and trustworthy businessman. Indeed, Muhammad's upright and dependable reputation earned him the designation al-Amin (the Trustworthy One) among his fellow Makkans, and even invited a marriage proposal from Khadijah, a businesswoman in Makkah for whom he worked.

At the age of twenty-five, Muhammad married Khadijah, a widow who was his elder by fifteen years. Their marriage lasted twenty-five years, until Khadijah's death. Muhammad and Khadijah had six children: two sons died in early childhood and four daughters lived to bless their household.

While most of his fellow Mekkans were polytheists, Muhammad refused to worship the traditional tribal deities and often retreated to meditate and worship the One God of his ancestor, Abraham.

At the age of forty, while meditating in the cave of Hijra in the mountains above Makkah, Muhammad received the first of many revelations, beginning with the Arabic word "Iqra", meaning "Read" or "Recite". Soon afterwards, he was commanded to convey the Divine message and thus became the last messenger of God, according to the Qur'an.

> Read, in the name of thy Lord, Who Created –
> Created man, out of clot (embryo).
> Proclaim! And the Lord is Most Bountiful,
> He Who taught the use of the pen -
> Taught man that which he knew not (96:1-5)

Muhammad spent the remaining twenty-three years of his life receiving revelations from God and advocating the message of Islam among the peoples of the Arabian Peninsula and working to implement the principles and teaching of Islam in human society. After suffering severe persecution from the polytheistic Mekkans for 11 years, he and his fellow Muslims emigrated to Yathrib, a city 200 miles north of Makkah, where he established Islamic rule. The city was renamed Madinah (short for Medunat an-Nabi), City of the Prophet. In the following years, the message of Islam brought more and more tribes in the Arabian Peninsula into the fold, creating a new community based upon common religious principles, rather than tribal or other affiliations.

Muhammad died in 632 C.E. at the age of 63. His tomb is located adjacent to the Masjid an-Nebensi in Medinah, Saudi Arabia, in what used to be his quarters next to the original masjid of the city. (Shaikh, 1995).

Aiyub Palmer

BAHA'I

The Baha'i Faith is a religion that was founded in Persia, now called Iran, in the middle of the nineteenth century. It teaches that all people of the world are members of one human family. It also teaches that all religions are basically the same, and that all religions worship the same God. According to the Baha'i Faith, one day all the countries of the world will join together as one (planet earth) with one (community of religions).

The founder of the Baha'i Faith was called Baha'u'llah. He taught that all the great religions of the world are divine in origin, that their basic principles are in complete harmony, that their aims and purposes are (ultimately) the same and that their teachings are aspects of one truth. He explained that a Messenger from God is sent about every one-thousand years to provide spiritual guidance to humankind in a continuous and progressive process. Their missions represent successive stages in the spiritual evolution of human society. The unity and oneness of humankind is the central point of the Baha'i Faith, and the point from which all of its other teachings grow.

The Baha'i Faith is a widely disseminated religion, and one which is growing rapidly. Its central religious figures are the Bab (1819-1850), called the forerunner; Baha'u'lla'h (1817-1892), the Prophet; Abdu'l-Bah'a (1844-1921), the Interpreter; and Shoghi Effendi, the Guardian of the Baha'i Faith.

L.P.S.

ABOUT THE SCRIPTURES

The Baha'i Scriptures are drawn largely but not completely from the writings of Baha'u'llah, the prophet of the Baha'i Faith. Other teachings are from Abdu'l Bah'a, the Interpreter, and Shoghi Effendi, the Defender of the Faith. They include the following teachings for which the Baha'i Faith is very well known:

* The oneness of humanity. All peoples are members of the human family, all have the same Creator.
* The independent investigation of truth. Baha'u'llah teaches that each person must investigate truth for him/herself - that faith in this day must be built on knowledge and based on one's own decision.
* Elimination of all prejudices. For example: gender, religious, age, nationality, racial, economic, appearance, cultural.
* Agreement of science and religion. Baha'u'llah teaches that true religion and true science are in complete harmony.
* World peace. This is the time when all nations must seek by every means in their power, to establish cooperation among all the nations of the world.
* Equality of men and women. Both men and women, are equal in the sight of God, and both have the same rights and responsibilities.
* Universal education. Everyone, rich and poor, men and women, should receive an education.
* Consultation. Truth and the best decisions emerge from a process of honest and open discussion by every member of the group.
* Limitation of wealth and poverty. Society must not permit extremes of either wealth or poverty. The economic problem is essentially a spiritual one.
* Universal language. Baha'u'llah recommended the adoption of a common second language for all peoples in order to promote greater understanding between nations and individuals.
* Unity in diversity. The different languages and cultures of the world should maintain their individuality, but there must be a common link between them that can bring about understanding.
* A world commonwealth of nations. This is the time when the peoples of the earth should meet as equals. Their governments will be represented in a world parliament that will be concerned with the prosperity of all nations and the happiness of humankind.
* Progressive revelation. God has unfolded divine truth in successive stages through the whole history of humankind. This means that religion is evolutionary - it develops and progresses over the ages. Baha'u'llah's mission is to help unite humankind into a global community.

Bridging our Faiths, *prepared by the Interreligious Council of San Diego, in conjunction with the National Council of Community and Justice. Published by Paulist Press, 1997.*

SCIENTOLOGY

In the introduction of his book, *Science of Survival*, L. Ron Hubbard acknowledges fifty thousand years of thinking scholars without whose speculations and observations the creation and construction of Dianetics (a branch of Scientology that addresses what the soul is doing to the body, derived from the Greek dia=through and nous=soul) would not have been possible. He gives specific credit to twenty-three philosophers of Western civilization and his instructors in atomic and molecular phenomena, mathematics, and humanities at George Washington University and at Princeton.

Scientology is the study and handling of the spirit in relationship to itself, universes and other life, codified and implemented into counseling procedures by the founder. Scientology offers a unique synthesis of scientific inquiry and religious philosophy, comprising traditional knowledge and modern discoveries and utilizing most recent developments in electronics. It is based on observation, a formulation of axioms, with predictable results in their technological applications; while at the same time, it gives perspective as to purpose and evaluation in respect to ethics. As a religion, Scientology lives in the tradition of the great religions of the East, closest related in thought to Buddhism. Scientology is not dogmatic in regard to a Supreme Being, but encourages its followers to be inquisitive and to discover truth for themselves. It lays out a road and provides the tools to salvation, which to a Scientologist is eternal spiritual freedom. Consequently, Scientology is non-denominational and open to everyone, independent of his or her religious background. In fact, Scientology incorporates the elements essential for a global religion that adequately serves all traits of ethnic and educational background.

ABOUT THE SCRIPTURES

In the Scientology Scriptures, within the *World Scriptures*, excerpts from Mr. Hubbard's writings are given from various Scientology scriptures; however, only a comprehensive study will enable the interested reader to acquire a real understanding of the philosophy and religious practice of Scientology. Born of his research in the field of study, the following advice may be given: "The only reason a person gives up a study or becomes confused or unable to learn is because he or she has gone past a word that was not understood". This factor of study is also why you will find some new terms in Scientology. They are not new names for elements already understood. Vital to an understanding of Scientology is the realization that many elements necessary to an understanding of man have not been accurately defined. These elements have been precisely observed and those observations are then passed on to the student for his own verification from his own viewpoint. A fundamental premise common to all of Mr. Hubbard's teachings is that a datum is only true for you if you have observed it for yourself and find it to be so.

Scientology does not rely on revelatory writings to impart understanding. Scientology is a religion of self-deliberation without reliance on exterior forces. Beyond the communication of its basic philosophy and principles, the religious scriptures are devoted to the education of its disciples to acquire the knowledge and skill to apply these principles.

When we study the works of L. Ron Hubbard, we immediately recognize that his primary interest serves the spiritual aspects of humankind, without reference to body forms. The words *man* or *men*, as well as use of *he*, *his*, or even the terms *fellow* or *brother* are consistently understood to refer to spiritual beings, irrespective of gender.

Wolfgang Keller

BRAHMA KUMARIS

In 1936, in a country that is now called Pakistan, a being describing himself as "The Supreme Soul Shiva" - the Knowledgful, Luminous, and Blissful Self" - entered the 60-year-old body of a well-established Hindhi gentleman, a multimillionaire diamond merchant named Dada Lekraj.

In a short while, several hundred people began to gather on a regular basis to listen to the teachings of Shiva being spoken through Dada, whom Shiva renamed Prajapita Brahma (The Father of Humanity). The community of these people eventually became known as the Brahma Kumaris. The teachings, referred to by Shiva as the Ancient Raja Yoga of Bharat, became a daily morning discourse known as the Murli, which, literally translated, means "flute".

ABOUT THE SCRIPTURES

Now, for the first time in the history of the Brahma Kumaris organization, a direct translation of some of the major points of the Murlis is being presented. The extracts have been grouped under two main subject areas of the Brahma Kumaris curriculum, and within that they have been ordered according to linear logic. To present the Murlis in their entirety in the non-linear fashion in which they were spoken, and with sufficient annotation to make them coherent to the yogicly untrained, would have been a daunting task. However, for anyone who wishes to complete a basic introductory course in meditation and spiritual knowledge provided without charge at any of the more than 5,000 Brahma Kumaris centers worldwide, the full Murlis are available for further study.

Sister Jayanti
Brahma Kumaris World Spiritual University

THE SUFISM OF BAWA MUHAIYADDEEN

Muhammad Raheem Bawa Muhaiyaddeen was a Sufi saint, a man of extraordinary wisdom and compassion. For over seventy years he selflessly shared his knowledge and experience with people of every race and religion, and from all walks of life.

Born well before 1900, Bawa Muhaiyaddeen spent his early years traveling through the Middle East and India, examining the world's religions and a myriad of spiritual practices. Our record of his life begins around 1914, when pilgrims traveling through the jungles of Sri Lanka first encountered him. Awed by the depth of his wisdom, they asked him to come to their village to be their teacher. Some time later he did so, thereby beginning a life of public service - feeding, healing, and uplifting the lives of all who came to him.

The name Muhaiyaddeen literally means "one who restores to life, faith and the path of purity". Indeed, Bawa did devote himself tirelessly to awakening faith within people's hearts. As a Sufi, he had that special gift of distilling and revealing the essential truth contained within all religions - the oneness of God. Thus wherever he spoke, Buddhists, Hindus, Muslims, Jews, and Christians would sit together for hours listening to his wisdom.

In 1971 he was invited to come to Philadelphia, the "City of Brotherly Love". There, learned and unlearned, young and old responded to his message of unity. Over the next fifteen-year-period until his death in Philadelphia in December, 1986, he divided his time between the United States and Sri Lanka. He spoke on university campuses, in churches, meeting houses, and private homes, as well as on numerous radio and television programs. He reached audiences around the globe, from the United States and Canada to England and Sri Lanka. He was interviewed by *Time*, *Psychology Today*, *Harvard Divinity School Bulletin*, *The Philadelphia Inquirer* and numerous other publications. Over twenty books of his discourses and songs were published, as well as scores of audio and videocassettes, and artwork.

ABOUT BAWA MUHAIYADDEEN'S TEACHINGS

Four teachings of Bawa Muhaiyaddeen stand out in his writings:
(1) The Meaning of a True Human Being
(2) The Mind
(3) The Shore of the Heart, and
(4) The Path of God.

(1) A true human being is one who manifests the One God existing in all hearts. We have a sixth level of consciousness called "Divine Wisdom" within us. This we need to cultivate as our principal reason for being here.

(2) The mind helps us to function in this world, but the essence of life comes from the soul, which is deathless. Be in touch with the soul and use your mind to deal with the world.

(3) The good people who have transcended the mind can reach the shore of the heart which then will lead them to God. Push back the words of the tongue and speak only God's words.

(4) The energy understood by the five senses, maya, is not the power which can explain. That power is the Original, Great Effulgence. This must be understood. May God protect us all.

Richard Kuznetsky, M.D.

NATIVE AMERICAN TRADITIONS

Dhyani Ywahoo is a member of the traditional Etowah Band of the Eastern Tsalagi (Cherokee) Nation. Trained by her grandparents, she represents the twenty-seventh generation to carry the ancestral wisdom of the Ywahoo linage. In 1969, after generations of secrecy, it was decided to share the teachings of the Tsalagi tradition with non-Native people, to help save the environment.

Charged with the duty to rekindle the fire of clear mind and right relationship in these changing times, she has become a guide to all who walk the Beauty Road. She is the founder of Sunray Meditation Society in Bristol, Vermont. The selections are taken from her books *Voices of Our Ancestors*: *Cherokee Teachings from the Wisdom Fire* (Boston, 1987).

L.P.S.

ABOUT THE SCRIPTURES

TSALAGIA ELO - our philosophy, our oral tradition - tells how the Principal People, the Ani Yun Wiwa, originated in the star system known as the Pleiades, whence first arose the spark of individuated mind.

The different sections of these teachings are: *The Peoples of the Fire; Tribe Activity/Energy; Three Fundamental Truths; The First Sacred Fire: Will, Intention to Be; The Second Sacred Fire: Affirmation, the Wisdom Energy of Compassion; The Third Sacred Fire: Actualization, Skillful Means; Affirmation; Vigil; Spiritual Practice; Ten Sacred Stones: The Rainbow Bridge*: and *Meditation on Clear Intention*.

Go through the day with peaceful thoughts, doing good things.

Dhyani Ywahoo

CAODAI

In the earliest days of human existence on earth, direct contact with the Creator was more common than it is today. The Creator spoke in ways and symbologies relevant to each culture, for each tribe had its own land, language, and customs distinct from the other. The Magnanimous Force we know as God loved not one more than the other and communicated directly, answering each in its own tongue. Humanity's numbers were yet few, spread sparsely, mostly isolated from one another. Glorious cultures and religions, all springing from God who had created them, developed everywhere independently. As time passed, humans more and more forgot how to communicate directly with God, though they still held dear their religions.

When the various tribes began to encounter one another upon the earth, they were mystified, not understanding why these others were so dissimilar, appearing and acting differently, worshiping differently, and surely each thought the other must revere a different God! Because humans did not understand one another, they had fear, and as children who are afraid and away from the guidance of a parent will become bullies, humans created friction and war. They did not recognize each other as children of the same Creator. Even though there were further revelations from God to those who could still hear (*Love your neighbor as yourself* - Christianity; *There is no God but God* - Islam), rather than embracing diversity, people still preferred to resist the differences between themselves and continued to make war with one another. Separations became more and more minute, until even a single principle was allowed to splinter the mightiest of religions into sect upon sect.

Then, in the last century of the second millennium, with the bodies of eight million victims of World War I still decomposing (a testimonial to the great evil of war), God spoke to an unrenowned esotericist practicing his self-cultivation (meditative arts), at a far corner of the earth, an as-yet unknown place we today call Vietnam.

Precisely, this contact occurred on the Vietnamese island of Phu Quoc, south of the mainland of Cambodia. There Ngo Van Chieu received messages directly from God relating to a great reconciliation which must take place between all religions. Chieu worshipped and vowed to obey God, Whom he knew by the name of Cao Dai. He asked if he might have permission to worship God in a representation he could view. God replied that the human form is too limiting to represent the essence of universality that God embodies, and that the Eye would be more apt, as it is the symbol for both Universal and Individual consciousness.

So the All-Seeing Eye of God (Thien Nhan, Divine Eye) became the symbol for CaoDai, for God, and for the ensuing religion which is called CaoDai.

> God said:
> The Eye represents the heart
> From which twin pure lights beam.
> Light and Spirit are One,
> God is the Spirit's gleam.

ABOUT THE SCRIPTURES

The CaoDai Scriptures stress the essential message that all religions are One. God, celebrated as Source of the Universe and all souls, and as the origin of all religions, is seen to have manifested differently in different epochs and be called by myriad names. CaoDai teaches human beings, who all have sprung from the same Source, to live in harmony, love, justice and peace; to enjoy universal sisterhood and brotherhood; and to cultivate themselves to seek and be reunited with God in their hearts.

That this flower of peace took seed in a land destined for war may seem a contradiction, but once understood, the CaoDai path simply transcends all contradictions and leads the world to tranquility. Within these teachings is found the revelation of a unique yet all-encompassing path to God.

Hum Bui, M.D.
in collaboration with
Ngasha Beck

SCIENCE OF MIND

Ernest Holmes, founder of the Religious Science movement, was born on a farm in Maine in 1887. He grew up in a poor but very literate family and was an especially curious youngster who read avidly but found school dull. Before finishing high school, he abandoned formal education and set out on a lifetime of independent study. Early on, he discovered Ralph Waldo Emerson, with whose teachings he immediately felt a deep identification. While living in Boston, Holmes was introduced to Christian Science and soon developed a mental healing technique similar to that practiced in Christian Science. In 1912 he moved to Los Angeles, and after coming across the works of leading New Thought writer Thomas Troward, Holmes began lecturing on metaphysics. Growing demand led him in 1927 to establish the Institute of Religious Science and Philosophy, which later became the Church of Religious Science. Its first formally organized congregation was Founder's Church in Los Angeles, which at one period had more than 10,000 members. By the times Holmes died in 1960 there were eighty five Religious Science churches nationwide, as well as hundreds of licensed Religious Science ministers and practitioners. Holmes was a student of all the world's religions and philosophies, and believed they were all pointing to the unity found in one ultimate, universal Truth.

93

Excerpts are taken from: Ernest Holmes, **Can We Talk to God?** *- Science of Mind Publishing, 1934, 1992.*

Kathy Juline, Editor
Science of Mind Magazine

ABOUT THE SCRIPTURES

The Science of Mind Scriptures contain many of the teachings of Ernest Holmes that are put forth by the Church of Religious Science. They include The Meaning of Freedom, Mental and Spiritual Laws, God and Human Beings, Nothing Happens by Chance, We are a Center of God-Consciousness, The Aim of Evolution, Methods of Treatment and Place no Limit on Principle.

More and more, the religions of the world will need to apply the discoveries of science having to do with consciousness and the newly discovered universe in which we live. Science of Mind has taken a most important step into this realm by striving to benefit from science and religion in seeking the meaning of life.

L.P.S.

CENTRAL SCRIPTURES

The "New Testament" of World Scriptures

World Scriptures is an effort to bring together the sacred teachings of our universal heritage from the beginning of recorded time to the present and beyond. If we are to have a sense of meaning for our lives in the emerging global civilization, it is imperative that we come to understand the world's religions and spiritual movements, both ancient and modern. Most of **World Scriptures** has this as its task.

However, as with the **Holy Bible** of Christianity, there must be a "New Testament" consisting of the essential teachings of the world that is emerging. This is the prophetic aspect of religion, which is essential to its fulfillment. These new materials in **World Scriptures** include writings from the science of consciousness, from the teachings of Sri Aurobindo, from Mahatma Gandhi and Dr. Martin Luther King, from numerous poets and seers such as Kahlil Gibran, and from some of my own writings. Prophecy in our age will not be confined to any one person, yet it is necessary that the powerful and diverse civilization that is just now beginning to become visible, will have its own guidance. ("Central Scriptures" and related sacred writings will be published in a forthcoming edition).

The *World Scriptures* would not be complete if it ignored this prophetic dimension. The birthing of a new conception of scripture is an unfolding process. It will continue to be formed for some years to come. Nevertheless, now is the time when the new scriptural guidance is being written and brought together. You are cordially invited to be part of this process in whatever way is right for you and for those who are helping it to happen.

In the Spirit of Unity-and-Diversity!!!

L.P.S.

AUSTRALIAN ABORIGINAL SPIRITUALITY

Reverence for Life

A *spirit child* selects its mother and enters into a woman's body. This is the story of birth in an ancient aboriginal culture. The mother may be a human being, a walrus or a kangaroo. The great family of the aboriginal is the earth with its rocks and stones, trees and mountains, and all the living creatures that reside on it and under it in the oceans and in the rivers and in the sky. The spirit is *reverence for life* that accepts the priorities of the food chain and kills only what can be consumed.

The Australian aboriginal has no written language. The oral singing traditions go back over thirty thousand years into the Dreamtime or 'in the beginning'. In the beginning was the song, and the song was with the great creator, and the song was the great creator. Though there have been over five hundred languages and many somewhat different mythologies, there is an unmistakable central theme to a spirituality that could be shown to be the most scientifically tenable on planet earth. The price paid for it is the non-existence in the culture of the levels of civilization that go beyond the social. Territories were not owned but inherited by birth. True freedom was experienced in living in one's own country, the country that one belongs to by birth, and where one does not have to ask permission to walk. Material things were minimal and expendable. Shelters were made from materials at hand and not left standing as a tribe moved on though some natural rock utensils would remain. Everything on the face of the earth had its own role to play so covetousness of personal qualities had no meaning. People took pleasure in developing their own selves. They would invite friends to celebrate, not birthdays, but spiritual progress at which time they may even take a new name to signify this personal progress. Songs told their story, songs created their identities, songs respected the rights of others' 'countries' and were their passports, songs and dances were their celebration of life, paintings on bark and later on canvas, were gifts to the world and in the grottos and sacred places, paintings recorded their history and their destiny. What matters is the essence of the understanding of the role of human beings in the workings of the universe, and how to live in harmony with it. Aborigines brought no harm to their environment and lived free, meaningful and fulfilling lives in a *spiritual sense* without the additions and benefits of civilization.

- To move, always as the animals move, gently across the earth, taking no more than what was needed, leaving no trace. To lie down on the earth at night time and watch the greatest movie ever shown, picking up each night from the night before, brilliant stars, many dropping; the clouds from white to grey to totally departing and the moon never letting you down, always returning, always revealing the beauty of its journey from the yellowness of its brilliant fullness to the soft glow of its waning light among the stars who never fail to reveal the harmony ever and always present.-

What is the purpose of life? We are all born, and to each one of us is given a certain number of days or years to spend on the face of the earth - doing what? There is a great warmth in understanding that we belong to the universe. To gaze across flat land as far as the eye can see, a few bumps in the distance and then the sky, ever changing, ever magnificent, from dawn till dark not a moment without change, not a moment without beauty. Each day in an arid land, prayers were offered to other creatures to give up one of their own for the sustenance of this tribe this coming day. Rewards came as thanks were given for whatever was provided. An animal taken was accepted with gratitude, and its spirit released or the nutritious witchetty grub was relished for its gift to the body. Those who secured them were rewarded daily with the joy of giving, of providing for the group whatever the earth would yield, of sharing it according to need. Everyone understood this method of governance, for it was written in spiritual law. Relative to the needs of all, the supplies were sought and thanks were given. Nothing was wasted, for whatever was unconsumed or not used for utensils, was returned to the earth.

This was the aboriginal environment. Do we have it better who live in cities and rarely see the skies at night, the clouds by day. Do we recognize for what it truly is, the sun, as it touches a pavement or moves across a tall building, a reassurance that the Great Spirit is fulfilling its duty on the earth and setting the example for us to follow, that all is well, that our lives are to be lived within the glow of the universe and guided by its ever present wisdom and continuity.

Can there be more material deprivation than in the life that the aborigines led? For over thirty thousand years they adjusted to the land's barrenness and lived in harmony with it. What great truths can the progress of science reveal in their ways?

Their gift to the twenty first century is the purity of the spirituality. It was not their responsibility to add the layers of civilization. It was their responsibility to keep pure what was in their charge and they did that. It is our responsibility now to endeavor to understand it, to retrieve its directives as we all struggle to re-achieve the harmony which is our birthright as well as theirs and which we are all now in danger of losing.

SONGS OF CREATION AND LAW.

Those who sing the songs do not compose them. They receive them often in the Dreamtime. They 'catch' the song or the dance or the painting and bring it back to the tribe. In the beginning was the creation time and the first being was the great Earth Mother, Eingana, who lived in the middle of the earth and rose up out of her pool of water which arose after flooding. She made everything. She made everything that flew or walked. She made all the rivers and the rocks. She made all the people and she made the Rainbow-Snake. She took them all back and again let them all out. They are all her spirit children, and she always holds the thread of their life. When the body dies, the spirit returns to her earth, the place where it was born or where it entered the woman. Only Eingana, the great Earth Mother, gives out spirits. If she ever died there would be nothing, no animals, birds, people, water, nothing.

One day the Rainbow Serpent was sitting in the middle of a river singing her sacred songs with her Larnja, the sacred source of all ritual, beside her. A tribal elder heard her, carefully sneaked up behind her and stole her Larnja. He gave the ritual and the laws to various tribes and the men have held them ever since. The men's initiation rites now include great pain but it is different from the pain of the woman after the spirit child has entered her body and decides to be born. The rituals and the laws that the Rainbow-Snake gave to her daughters, have been kept in their hearts and secret from that time. Even though the men stole the sacred ritual of the ancestral creation-time spirits from the fertility goddesses, they still stand in awe of the great earth mother; and they use her laws to organize their societies. The tribal elders are simply the custodians of it. .

M.A.M.

THE SOUL OF AFRICA

Long before the great religions recorded their scriptures, the vast continent of Africa was inhabited by various peoples with extensive mythologies, many of whom recognized a Supreme Being whose home was in the sky and in the afterworld of the dead. Ancestor worship was prevalent. In some instances, the spirits of the ancestors were believed to guide the destinies of their living descendants.

Many African people also make considerable use of magic and have faith in omens of various kinds.

Because of the great inroads made by invading armies, and the introduction of various religions including Islam and Christianity, it is not as easy to characterize a spirituality for the African people as it is possible to do for the comparatively untouched inhabitants of Australia before their European influx in the nineteenth century.

However, in 1966, during the Black Freedom Movement in America, a cultural phenomenon of African identity was recognized and recorded by Dr. Maulana Karenga, and a pan-African holiday began to be celebrated throughout the world African community. It is called **Kwanzaa**, a celebration of family, community, and culture. While it has been stressed from the beginning that this is a cultural rather than a religious celebration, there is a spirituality in **Kwanzaa** that makes it an integral part of any book on universal spirituality.

It is estimated that there are now fifteen million people around the world who celebrate Kwanzaa.

Kwanzaa brings a cultural message which speaks to the best of what it means to be African and human in the fullest sense. It represents an ancient and living cultural tradition that reflects the best of African thought and practice in its reaffirmation of the dignity of the human person. It includes community and culture, the well-being of family, the integrity of the environment, and our kinship with it, and the rich resources and meaning of a people's culture.

The name *Kwanzaa* means "first fruits" in Swahili, a Pan-African language which is the most widely spoken African language. These first fruit celebrations involve: ingathering which re-affirms bonds among people, reverence which recognizes the Supreme Being or the Creator and gives thanks for the bounty and the beauty of all life, commemoration which honors the spirits of ancestors and those human beings who have excelled as models of human endeavor, re-commitment which involves renewed dedication to the highest cultural ideals of the African peoples and celebration of the Good which reaches the Good of the Divinity and raises this cultural phenomenon to the spirituality where it takes its place among the great religions and other spiritual movements that are moving our global civilization into the 21st century.

Kwanzaa brings African culture back to its roots and reinforces seven ancient communitarian values of Unity, Self-Determination, Collective Work and Responsibility, Cooperative Economics, Purpose, Creativity, and Faith.

The celebration of *Kwanzaa*, a cultural holiday, begins on December 26 and ends of on January 1. The ceremony involves the symbols of the Crops, the Mat, the Candle-Holder, the Corn, the Seven Candles, the Unity Cup, the Gifts and the Flag whose colors are black, red and green - black for the people, red for their struggle, and green for the future.

The last day, January 1, is a day of meditation and assessment of the ageless question 'Who am I? Am I who I say I am and am I all that I can be?'

a cultural holiday with a very spiritual message

Kwanzaa material adapted from the Kwanzaa website

M.A.M.

NATIVE AMERICAN PEACE EXAMPLES

All living things are tied together with a common navel cord. (Sioux)

Though there are many traditions of many tribes, hundreds of languages, and thousands of dialects, Native Americans as a people can be understood as a microcosm of a spiritual world that science can affirm.

As a people, American Indians have a holistic view of the world, or a sense of an integrated universe with a mysterious Great Spirit (which has many names as *Manitou* or *Wakan Tanka* or *Orenda*) that is accessible to all animate and inanimate presences in the universe. There is no difference between the sacred and the secular. Everything that can be seen or touched, animate or inanimate, has a spirit. The earth is sacred, and all living creatures which are sacrificed for human sustenance are honored. The society is egalitarian. In fact, "all my relations" includes the equal partners of man, woman, plant, animals, including everything that crawls, runs or flies as well as stones. As all of the parts contribute to the well being of all, punishment is an alien concept. The ruling power is the Good, the good for all. Children learn by example. There are no jails, no homeless and no religious animosities.

The various tribes, however, have their own distinct deities, powers, rituals, and story-tellers who passed on from generation to generation, various tales of creation and tribal journeys. Though the tales, as the journeys, may vary, the destiny is one destiny as it is for all life.

The Mystery and the Power of the Life Force, or the Great Spirit of the Universe.
Conscience, common sense and natural law.

At the point of the first breath taken, an individual is born, a conscience is born, of a womb, of the womb of the great Earth Mother, of the great Spirit of the Universe, of the Life Force. The point of an individual birth is a point on the hoop or the circle of a family, a community, a society, a culture, an environment, a planet, a cosmos, where the beginning is a beginning of a journey along the circle that ends for each individual where it began, in the womb of the great Earth Mother the first drawer of the breath of life from the Life Force or the Great Spirit of the Universe.

Keeping a pure or a good conscience is common sense for the American Indian, governed by natural law. It entails honoring and respecting Mother Earth. From her springs plant life to eat and to use for healing; from her emerges the pure water without which life withers and dies; buried deep in her body are deposits of coal and uranium and oil shale and ancestors which are all sacred to her and the great Spirit of the Universe or the Life Force; from her womb come the wombs of the global mothers and the grandmothers; these wombs have the *right* to be productive, to see their children rise up to and beyond the seven generations - and not be cut down by war machines made from coal and not denied the pure rain that comes from the lightening that comes from the uranium - and not denied the right to hunt the sacred animal for their sustenance, to fish in pure rivers - not deprived of their freedom to live on their own land and raise their children in the path of the pure conscience.

The journey of the pure conscience accepts responsibility for its own actions but now it must also accept responsibility for alerting the global conscience to the damage already done to the earth and to the environment so that destructive practices may be halted and then reversed.

Wisdom Keepers are raising their voices as they have never raised them before. They are speaking at the United Nations, they are publishing stories explaining their own journey along the path of the pure conscience, they are setting examples for others to gaze at and attain their own vision of true spirituality, the knowledge of the individual conscience that each one of us must accept responsibility for our own actions, now to be directed toward reversing the damage, turning away from destruction and preserving the planet for our own children for seven generations whose little faces even now are pressing against the womb of Mother Earth pleading for their birthright.

If we might like to ascribe something resembling a religion to the American Indian People, and an unfulfilled gift to the colonists who took control of their land, it might be called the religion of the Great Peace guided by the historical and legendary Hiawatha who lived in the eastern woodlands in the fifteenth Century. This is an inspirational story for all the world and different from the Native American Church, which was what the Native Americans were able to secure for themselves in the face of great efforts to annihilate them.

In the religion of the Great Peace, the symbolic core or center is Mother Earth infused with *orenda* or psychic energy of the Great Spirit. In the Iroquois Confederacy, long before the coming of Columbus to America, the clan mothers were considered the progenitors of the nation and were granted ownership of the land and the soil, for it was the mother of the world who gave birth to all the messengers. Because of this sense of world motherhood and the sense of the human family dependent upon it, killing was not considered a worthy activity for those who symbolically sprang from the womb of the same mother.

The killing of human by human was in the peaceful process, followed by wampum or spiritual compensation to the wronged. To follow a killing with another killing was understood to be self-perpetuating atrocity only completed with the death of all.

This maternal understanding given to Hiawatha by his mother led to the great symbol of enduring democracy, the Tree of Peace which "pierces the sky" for the whole world to see, whose roots extend to the farthest parts of the earth, whose sustenance is aided by all the weapons of war buried beneath it, and whose longevity is assured by the ever-present eagle hovering above it, watching out for any approaching dangers.

This great democracy valued consensus in decision making. To guarantee this consensus, clan mothers - the givers of life to their men, and who were the owners and holders of the soil - were also given the civic responsibility of nominating political and religious leaders and monitoring their performance. If their leader sons failed to serve the best interests of all their people, tried to look toward war, or do any other acts which could harm the interests of their children and their grandchildren through seven generations, Clan mothers and women's councils had the right, after three warnings, to impeach them, to take back the emblem and title of Statesmanship and to confer it upon another of their sons.

Greatly indebted to the Iroquois Confederacy,
when the American Constitution was drawn up,
the need for full participation of mothers and grandmothers
in order to recognize the necessity for peace, and guarantee it,
was ignored.

Women were denied full participation in government in the United States,
not even securing the vote until the 20th century
and the
Great Peace
envisaged by Hiawatha and those who bore him,
has yet to be realized.

The Eastern Tsalagi, formerly Cherokee, shares much of the history of the Iroquois Confederacy.

M.A.M.

THE SPIRITUALITY OF GREAT LITERATURE,

THE WRITINGS OF THE GREAT PHILOSOPHERS, SCIENTISTS AND ARTISTS

Great literature expresses the human spirit in such a way that every age gains sustenance from it.

Artists are born with a passion to create the beautiful, to find the thread of the Life Force, the Spirit, or the Breath of Life, and spin it off in its own creation, a work of art.

It is in great literature that the human spirit is freed from any constraints of dogma or beliefs that come into play as a person endeavors to abide by or live within the parameters of any prescribed or organized religion.

For the sake of clarification, we may say that there is a communal gene or thread of life that runs through every living creature as well as objects such as rocks, formerly called inanimate, so that everything on the face of the earth, above it, and under it , partakes in this communal organism, the Life Force of the universe itself, the Spirit of All Life.

It is the function of great literature to clarify this. Samuel Taylor Coleridge calls it the imagination, the unfettered flight into the grandeur of consciously participating in the divine process of living itself. Shakespeare in his great plays, particularly his *tragedies*, reveals it to us with a clarity that is timeless.

The Oxford English Dictionary defines Philosophy as the love, study, or pursuit of wisdom. Wisdom is available to all human beings through their connection to the Life Force. Pre-Socratic philosophy led to the discovery of nature and hence Science. The appearance of what became known as Western science began with Thales, who posited that the universe was a natural whole with ways of its own which are beyond human control and the philosopher Parmenidies spoke of one world which contained the past, present and future. Socrates began the investigation into the relationship between the human being and God, the morality of aspiration, the meaning of virtue.

The very great philosophers describe their findings. This is particularly true of Socrates and Aristotle. In between these two is another great philosopher, Plato, who brought directives which have had a lasting impact on Western thought including Christianity.

All people are a product of their own age and however enlightened they may grow to be, there will still be some aspects of their thinking that retain their contemporary attitudes. The Oxford English Dictionary refers to Aristotle as "The Philosopher" and he may still be the greatest of all philosophers because of his work on the Science of Being Altogether. It is Aristotle who endeavored to take all aspects of existence and describe them in an all-inclusive way. His Prime Mover over two thousand years has evolved into the Life Force, the Spirit, the one God known by any name or no name. The essence of his work lives today in the reach of this grand endeavor. It is not diminished by the flotsum and jetson of his age as the condoning of slavery and the non-recognition of the value of women's contributions in all areas of human endeavor.

Of the philosophers who have followed him, some have had a greater impact than others in the area of 'right living' or 'wisdom' or 'virtue', but most still return to him for guidance in an insight, if not a theory, of everything.

Science as it has been understood and is recently defined, tends to be considered empirical. There is a measure of hubris in this which has restricted the so-called domain of science to human verification. Great scientists in the past have intuited a higher force of power, among them being Einstein's "spirit" in the beginning quotation to this book. Now in the emerging twenty-first century there are many scientists who believe that there is scientifically verifiable evidence of the existence of a Life Force of the universe in which everything participates. It is in a sense the closing of the circle of an advanced civilization and the universe of ancient peoples such as the Australian aboriginal.

Going beyond these previous intuitions of this indescribable but ever-present "force" or "spirit" until they, together, become capable of describing our universe, our theory of everything, our zero point field, our collective unconscious, our Life Force or our God, described by any name or no name, ALL disciplines have been converging to reaching "harmony" which unifies science, religion and spirituality, and which philosophy is now charged with describing, in the spirit of the human aspirations first recorded in our annals by the great Greeks, Socrates and Aristotle and company.

M.A.M.

THE "UNIVERSE" OF MONEY

Faith in the power of technology. Christian Delacampagne

*Only after the last tree has been cut down: only after the
last river has been poisoned: only after the last fish has been caught:
only then will you find that money cannot be eaten.*
Cree Indian Prophecy

Voluntary submission to the will of God is perhaps the most profound directive of all the religions, the point at which they all concur. To date, we have not clarified the meaning of the will of God. Most of the great wars of the last two thousand years have been fought with each side firmly believing they had God on their side. Dichotomous relationships among people have also not been clarified, and now a so-called third 'god' has been allowed to emerge. This 'god', the god of money, has been noted through the ages, as it has been intertwined with commerce or economies. "Give to Caesar that which is Caesar's and to God that which is God's". The inference in the past has been that the god of money represents temptation only, the temptation to take a portion of commerce or the economy and spin it out of the field of accountability and responsibility, and allow it to go into its own orbit, and create its own 'universe', taking from the many to benefit the few. Now however the god of money has become the definer of a civilization. It has become the maker of decisions that affect millions of people. Since, in the articles in this book, among others, this god may be shown to be not only a-moral but also un-scientific - that is, not responsible to the Life Force of the universe, not in harmony with the workings of the universe - unexamined dedication to it by those people in a position to influence the lives of others, is causing global fiscal chaos. This will not diminish until money is understood in terms of its relationship to the organic world and adjustments are made to bring it into harmony.

Let us attempt to define the relationship between an economy which has dependence upon commerce and that money which has the potential of going into its own orbit and orbiting its own god.

There is a vast difference between an economy and irresponsible dedication to, and pursuit of, money.

HUMANITY AND COMMERCE

Economy is the basis of society. When the economy is stable, society develops.
The ideal economy combines the spiritual and material, and the best commodities to trade in are sincerity and love.
Morihei Ueshiba, **The Art of Peace**

All things are made up of numbers. Pythagoras.

Commerce in the language of 'numbers'.

Commerce may be crudely represented as a logical equation or transaction:

human (may be an individual, a partnership or a corporation)
+ (research and development, where applicable)* + services/product
= money + profit = price

$$a + b = m = v + p = z$$

a. the human motivator or instigator of the transaction.
b. services/product
m. money, a medium of exchange (ideally, or formerly, a measure of value)
v. value
p. profit
z. price

*very large corporations as oil conglomerates, pharmaceuticals may have a large outlay for research and development before the product is obtained. This is then amortized or spread out over a determined period of time and the price for that period of time reflects the additional cost. Sometimes where humanity may not be uttermost, these corporations may continue to build this so-called start-up money into the price even after the reimbursement for it has already been completely amortized or completely paid off.

At the point of choice a person chooses from the innermost core. When a person walks into a flaming building to save a child, humanity prevails. When a person jumps into a river to save a drowning girl, humanity prevails. When a person builds a hospital where a community has not had services, humanity prevails. That choice is easy for us to understand.

When a person enters into a transaction, particularly when a person becomes part of a corporation, the innermost core of humanity may appear to be peripheral or even extraneous but such an appearance is unable to mask humanity's inherent role as integral to the ongoing movement of the organic universe, and hence its inherent role in *every* transaction.

It is scientifically untenable to engage in commerce with a profit-only motive which removes the transaction from an organic connection either to humanity or to the environment. Such a transaction is not only amoral but it attempts to prevail against the relentless drive of the organic unifier.

Since all profit-only transactions are originated by humans, those humans who engage in this activity exhibit hubris. The degree of the hubris is the extent to which the needs of humanity or the environment have been violated; humanity represents human beings who have been short-changed of benefits that are naturally theirs, such as clean air and water and the right to exist as human beings; the environment is represented by the determination in good faith of its highest and best use. In addition, these profit-only transactions are structured so as to be inaccessible to others. Profits are taken and transactions closed, the wealth irrevocably attached to a few and guarded as a medieval fortress while the peasants died in the fields. These entire transactions, just as those of feudal lords, lack the availability of verification or affirmation. This is unscientific in the annals of human-beingness.

Whereas the concept of wealth in earlier times was not considered necessary to spiritual well-being, since the importance of money has risen in hearts and minds and gone into its own orbit of irresponsibility toward the organic unifier, it participates in the hubris which will be called to account at the choosing of the organic unifier itself.

Organically, each person's service is related to time or the number of hours per day devoted to it. A particular person for those hours may be cleaning toilets or may be flying an aircraft responsible for the safety of others, or may be the head of a large corporation as discussed above, or may be the leader of a country responsible for the karma or sense of well being of an entire country or region. The quality of life of a community hinges upon the buying power of the money paid for what may be called the 'least' of the services. The spirituality inherent in this has also been noted through the ages. "Whatsoever you *pay* to the least of these my kinspeople, you pay also to me". In any business entity that is a partnership, company or corporation which controls the services of individuals there needs to be a justifiable ratio between the highest and the least paid human being contributing their services. It is scientifically untenable to have compensation for any person's services below the poverty level if any other person is being compensated thousands of times more than the poverty level. It reveals that the god of money is driving the organization or the country instead of the organic unifier. Where there is a price paid for a product, as distinguished from services or where the product is the ultimate outcome, that product conceived and produced by human beings cannot escape the requirements of being in harmony with the organic unifier, being scientifically responsible to it.

In another article in this series, wilful destruction or mutilation of the environment is shown to be unscientific. Technology aimed at wilful destruction is generated by the god of money and is orbiting in its own 'universe'. As soon as human beings understand this on the pulses, they will return to their own conscience, their own inner connection to the Life Force, the Spirit of the Universe, they will reach out to other members of the global community seeking the *commonality* known by any name or no name. They will understand that focusing or placing faith in the power and the products of technology that wilfully destroy humanity and the environment is scientifically and spiritually untenable.

They will finally melt their guns and their armored vehicles, and cruise missiles into oxygen to help to preserve the life of our one and only planet earth, and they will never use the power of the god of money to try to destroy her any more.

M.A.M.

HOW DO WE EXPERIENCE OUR OWN SPIRITUALITY?

Alone,

We pause and hold our entire being in readiness. We seek stillness. We know that we belong to ourselves, to the earth, to our God, known by any name or no name, to the Life Force of our common origin with the entire cosmos. We allow ourselves to feel our connection back through our umbilical cord, through our mother's umbilical cord, to great Mother Earth, to the Life Force, the God of our common beingness where we belong to the consciousness of everything. We move from the depths of our soul's origin on to the wavelength of the Eternal Infinity, the One. We attain what heights we can, knowing that we are journeying home.

In our group,

As we feel the warmth and the comfort of standing or kneeling or sitting among those who share our path, whichever path it be, among the world's religions, spiritual groups, independent seekers, we strive to feel our unity with the universal spirit, with all people and with all life forms.

We open ourselves to receive the human spiritual forces of life, peace and joy. We gaze into the eyes of another and feel their surging life force invigorating and energizing our own.

We endeavor to radiate these energies to the world.
By approaching every other inhabitant of planet earth with a
feeling of loving-kindness,
by gazing into their eyes and recognizing
the divinity there,
by acting accordingly,
we partake of the PEACE that passes understanding.

CIVILIZATION TIMELINE

This civilization timeline represents gleanings of human endeavor both creative and destructive from the building of the pyramids to the present time. It has been included for several reasons: to give us all a sense of wonder at humanity's achievements; to show us the chronological origins of the great religions and other spiritual movements; to place the lives of people who have influenced our civilization into an encouraging chronicle of human thought; to record events that have shaped the destinies of so many peoples, cultures and countries; to reveal patterns of destructive behavior that have persisted to the present day; to encourage us all and spur us all into action to make the twenty-first century into an epiphany of human endeavor where old patterns are recognized in their nakedness and addressed with spiritual fortitude and scientific confidence.

Note: B.C.E.   = Before Common Era (before Christ)
       C.E.   = Common Era (formerly Anno Domini)

We also ask for your indulgence in judging the merits of items included and omitted, suggesting that you take the overall picture and view it perhaps as a work of art.

B.C.E.

3100-2965	Pyramids built		Egypt
2870	First known settlement at Troy		
1800	Civilization in Palestine		
1582?	Foundation of Athens		
1500-1000	Vedic period		India
1313?	Foundation of Thebes by Cadmus		
1250-1183	Age of Homeric heroes		
1200	Moses	JUDAISM	Egypt
1192-1183	Siege of Troy		
1000- 500	*Upanishads,*	BRAHMANISM	India
900	Maories arrive		New Zealand
811	Buddhist temple of Borobudur		Java
800 - 600	Upanishads	HINDUISM	India
- 600?	Sappho		Greece
660 - 583	Zarathustra	ZOROASTRIANISM	Persia
640 - 546	Thales		Greece
604 - 517	Lao Tzu	TAOISM	China
550?	Sun Tzu	*The Art of War*	
599 - 527	Mahavira	JAINISM	India
560 - 480	Gautama	BUDDHISM	India
551 - 479	Confucius	CONFUCIANISM	China
530?	Bodhidhama	BUDDHISM (CHINA)	
500 - 250	*Bhagavad-Gita*		India
496?-406?	Sophocles		Greece
470?-399	Socrates		
427?-347	Plato		Greece
408	Completion of The Parthenon		Greece
399 - 300	Indigenous	SHINTOISM	Japan
386 - 358		BUDDHISM (JAPAN)	
384 - 322	Aristotle		Greece
221	The Great Wall of China built		China
106 - 43	Cicero		Rome

43 - 17?	Ovid	Rome
5 C.B.E.-29 C.E	Jesus of Nazareth CHRISTIANITY	Palestine
4 C.B.E.- 65	Seneca	Greece
? - 90	Epictetus	Greece
100	Christian church arrives	Africa
161 - 180	Marcus Aurelius	Rome
272 - 336	Flavius Valevius Constantinus	
330	Constantinople founded	
395	Hypatia 'The Philosopher'', teaching scientific rationalism	Alexandria
354 - 430	Augustine, Saint	
405	St. Patrick consecrated Bishop of Ireland	
414	Latin version of the Hebrew Bible	Bethlehem
570 - 632	Mohammed ISLAM	Persia
597	St. Augustine's mission to Kent	
600 - 699	Arab conquest	Africa
610	Prophet Mohammed receives first Revelations of the Quran	
622	Migration to Medina from Mecca	
628	Mohammed's peace initiative	
638	Fall of Jerusalem to Moslems	
638 - 713	Hui-Neng ZEN	China
640	Muslims conquer Palestine	
639 - 41	Muslims control Egypt and Syria	
644 - 650	Muslims conquer Cyores, Tripoli Rule Iran, Afghanistan	
661 - 750	SUFISM	Persia
691	Dome of the Rock completed in Jerusalem	
700	Maya civilization reaches new heights	Central America
740	*Beowulf*	
742 - 814	Charlemagne	
762	Foundation of Baghdad	
800 - 877	John Du Scot	
885	Armenia declares independence from Arabs	

900	Sheherezade tells the *1001 Nights*	
935 - 972	Hroswitha of Gandershim, Saxon canoness, playwright	Germany
1000	Polynesians arrive in New Zealand	
1050?	Normans abolish slavery in England	
1066	Battle of Hastings	
1077 - 1166	Abd-al-Qadar al Jilani	
1086	Doomsday Book	
1096 - 1090	First Crusade	
1099	Crusaders conquer Jerusalem	
1126?	*Rubaiyat* of Omar Khayyam	
1147	Second Crusade	
1175	Toltec civilization in decline	Central America
1189 - 1199	Third Crusade	
1207-1273	Jalal ad-Din, Rumi DERVISH	Persia
1215	Magna Carta	
1225?-1274	Aquinas, Saint Thomas	Italy
1258	Destruction of Baghdad by Mongols	
1265-1321	Dante, Alighieri	Italy
1338 - 1453	Hundred Years' War	
1340 - 1400	Chaucer, Geoffrey	England
1381	Peasants' Revolt	
1415	Battle of Agincourt	
1420	England and France declare 'Perpetual Peace'	
1430 - 1431	Capture and execution of Joan of Arc	
1441	Beginning of African slave trade	
	Portuguese exploration in Africa	
1453	Ottomans conquer Constantinople	
1455 - 1485	The Wars of the Roses	
1456	The Gutenberg Bible	
1466 - 1536	Erasmus	
1469 - 1527	Machiavelli	
1475-1564	Michelangelo	Italy

1483 - 1546	Martin Luther	
1485	Caxton's English Printing Press	
1490?	Christian missionaries	Africa
1491	Greek taught at Oxford	
1492	Christopher Columbus discovers America	
1497 - 1498	Vasco da Gama reaches India by the Cape of Good Hope	
1513	Portugese traders reach South China	
1517	Ottomans conquer Egypt and Syria	
1519 - 1522	Circumnavigation of the earth by Magellan's fleet	
1533	Separation of the English from Rome	
1526	First Printed New Testament in English at Worms	
1534	Act of Supremacy (Henry VIII head of the Church in England)	
1547-1616	Miguel de Cervantes	Spain
1549	First Book of Common Prayer	
1562-1635	Lope de Vega	
1564-1616	William Shakespeare	England
1588	Defeat of the Spanish Armada	
1590s	Dutch begin to trade in India	
1601	Torres discovers Australia	
1607	First settlement in Virginia	
1608-1674	Milton, John	England
1611	King James Bible	
1618 - 1648	Thirty Years' War	
1619	First slaves delivered to America	Virginia
	Birth of the newspaper in Europe	
1620	Pilgrim Fathers establish Plymouth	
1622	Marie le Jars Gournay publishes *Equality of Men and Women*	France
1623-1662	Pascal, Blaise	France
1630	Puritan settlement of Massachusetts Bay	
1637	Massacre of Pequot village	Connecticut
1653	Taj Mahal built	India
1660	English Restoration	

1662	Final version of the Prayer Book	
1665	Great Plague	
1678	John Bunyan's *Pilgrim's Progress*	
1688-1772	Swedenborg, Emanuel	Sweden
1689	Bill of Rights	
1694-1778	Voltaire	France
1704	Capture of Gibraltar	
1706-1790	Franklin, Benjamin	USA
1724-1804	Kant, Immanuel	Germany
1743-1826	Jefferson, Thomas	Virginia
1749-1832	Goethe, Johann Wolfgang von	Germany
1752	Gregorian Calendar adopted	
1756	Black Hole of Calcutta	
1756-1791	Mozart, Amadaeus Wolfgang	Austria
1764	Spinning Jenny invented by James Hargreaves	
1765	Invention of the steam engine by James Watt	
1770-1827	Beethoven, Ludwig von	Germany
1770-1950	Wordsworth, William	England
1775-1783	American Revolution	USA
1776	Declaration of Independence	
1782	Beginning of Sunday School (Robert Parkes)	
1782-1852	Webster, Daniel	USA
1789	French Revolution	
1791	Creation of Semaphore system of Commerce (Claude Chappe)	Paris
1793	First protestant missionaries in India	
1794	Slavery abolished in all French territories	
1798 - 1801	Napoleon occupies Egypt	
1803-1882	Emerson, Ralph Waldo	USA.
1805-1844	Smith, Joseph MORMON	USA
1809-1865	Lincoln, Abraham	USA
1809-1849	Poe, Edgar Allen	USA
1817-1862	Thoreau, Henry David	USA
1817-1892	Baha-u-Llah BAHA'I	Persia

1818-1883	Karl Marx	Russia
1819-1900	Ruskin, John	England
1821-1910	Eddy, Mary Baker CHRISTIAN SCIENCE	USA
1828-1910	Tolstoy, Leo	Russia
1830	France occupies Algeria	
1831-1891	Blavatsky, Helena THEOSOPHY	Russian Empire
1834	Emancipation Day. Slavery abolished in British colonies	
1835	First protestant mission	Samoa
1836-1886	Ramakrishna	India
1840	Britain claims New Zealand as a colony	
1840	Canada granted independence	
	Stamps devised as payment for mail	
1844-1900	Nietzsche, Friedrich Wilhelm	Germany
1847-1933	Besant, Annie	London
1848	Mexico cedes Texas and California to the United States	
	Women's convention demands universal suffrage	Seneca Falls
1854-1856	The Crimean War	
1856-1939	Freud, Sigmund	Austria
1857-1858	Indian Mutiny against British rule	
1858-1919	Roosevelt, Theodore	USA
1858	**Secret ballot introduced in South Australia**	
1861-1941	Tagore, Rabindranath	India
1861-1947	Whitehead, Alfred George	England
1863-1902	Vivekananda VEDANTA	India
1867	Invention of the typewriter	
1869-1959	Wright, Frank Lloyd	USA
1869-1946	Gandhi, Mahatma	India
1869-1940	Goldman, Emma	Russia
1869	National Woman Suffrage Association in America	
	John Stuart Mills, *The Subjection of Women*	
	Opening of the Suez Canal	

1871	Alsace-Lorraine annexed by Germany, Collmar the only part not ceded	
1871	Auguste Bartholdi, sculptor of Colmar arrives in the United States	
1872-1949	Gurdjieff, George I.	Turkey
1872-1970	Russell, Bertrand	
1874-1965	Churchill, Sir Winston	England
1875-1961	Jung, Carl G.	Switzerland
1875-1965	Schweitzer, Albert	Germany
1876-1969	Raj, Dada Lekh	
1877-1962	Hesse, Herman	Germany
1878-1965	Buber, Martin	Austria
1879-1855	Einstein, Albert	Germany
1880-1949	Bailey, Alice	England
1880-1931	Gibran, Kahlil	Lebanon
1881	France occupies Tunisia	
1881-1955	De Chardin, Pierre Teilhard	France
1882-1945	Roosevelt, Franklin D.	USA
1884	Auguste Bartholdi's *Statue of Liberty* presented by French Government to America	
1884	**Australian secret ballot system extensively adopted in The United States**	
1884-19..	Roosevelt, Eleanor	USA
1887 - 1960	Ernest Holmes SCIENCE OF MIND	USA
1888	End of African slave trade	
1889-1977	Chaplin, Charles	England
1890-1960	Pasternak, Boris Leonidovich	Russia
1890?-1986	Muhammad, Raheem BAWA MUHAI- YADDEEN	India
1892	Britain occupies the Sudan	
1893-1952	Yogananda, Paramhansa	India
1893	**Women gained the right to vote**	**New Zealand**
1894	Women gained the right to vote	South Australia

1894	Massacre of Armenian revolutionaries against Ottoman rule	
1895-1986	Krishnamurti, Jiddu	India
1895-1983	Fuller, Buckminster	USA
1897	First Zionist Conference in Basel	
1898-1979	Marcuse, Herbert	
1898-1993	Peale, Norman Vincent	USA
1899	Women gained the right to vote	Western Australia
1900-1980	Fromm, Eric	Germany
1902-1974	Lindberg, Charles	USA
1901	Foundation of Commonwealth	Australia
1901	Oil discovered in Iran, concession to British	
1902	Women gained the right to vote	Australia
1903-2003	Hope, Bob	England
1903	First flight by Orville Wright	North Carolina
	Emmeline Pankhurst forms Women's Social and Political Union	Manchester
1904-1988	Campbell, Joseph	USA
1904	Trans-Siberian Railway	Russia
1905-1961	Hammarskjold, Dag	Sweden
1906-1975	Arendt, Hannah	
1906	Women gained the right to vote	Finland
	San Francisco earthquake	
1908-1970	Maslow, Abraham	USA
1908-1986	De Beauvoir, Simone	France
1911-1986	L. Ron Hubbard SCIENTOLOGY	USA
1911	Roald Amundsen reaches the South Pole	
1912	Sinking of the Titanic	
1913	Women gained the right to vote	Norway
1914	Archduke of Austria assassinated at Serajevo	Austria
	Austria declares war on Serbia	
	Germany invades Belgium	

	Great Britain declares war on Germany	
1914-1919	**WORLD WAR I** *The War to End All Wars**	
1917-1963	Kennedy, John Fitzgerald	USA
1917	United States joins the allies	
1917	Balfour Declaration gives British support for Jewish Palestine	
1919	Treaty of Versailles	France
	Alsace-Lorraine returned to France	
1919-1921	Turkish War of Independence	
1920-1922	Gandhi organizes civil disobedience campaign against British rule	India
1920	Formation of the League of Nations	
1920	**19th Amendment gives women the right to vote**	**USA**
1921	Reza Khan founds Pahlavi dynasty	Iran
1921-19..	Glenn, John	USA
1922	Independence	Egypt
1923-1998	Shepard, Alan	USA
1923	Earthquake in Tokyo	Japan
1927	Birth of television	
	First talking film	
1927	Charles Lindbergh crosses the Atlantic	
1927	Flying doctor service established	Queensland
1928	Women over 21 obtained the right to vote	England
1929-1968	King, Martin Luther Jr.	USA
1929	First woman cabinet minister	London
1930-19..	Armstrong, Neil	USA
1930	Amy Johnson flies from England to Australia	Darwin
1936	Beginning of BRAHMA KUMARIS	Pakistan
1936	Civil war	Spain
1930? -	Ngo Van Chieu CAODAI	Cambodia
1932	Kingdom founded	Saudi Arabia
1938	Crystal Night	Germany
1939	Germany invades Poland	

	Great Britain declares war on Germany	
1939-1945	WORLD WAR II	
1940	Alsace-Lorraine again annexed by Germany	
1940	Evacuation of Dunkirk	
1941	Japan bombs Pearl Harbor	
1942	Surrender of Singapore	
1942	Morihei Ueshiba *The Art of Peace*	Japan
1942	Battle of the Coral Sea and Midway	
1944	Muhammed Reza succeeds Reza Shah	Iran
1945	ATOM BOMB on Hiroshima and Nagasaki	Japan
	Surrender of Japan	
	Alsace-Lorraine returned to France	
1945	**Founding of the United Nations**	
	Turkey joins the United Nations	
1947	Formation of the Arab League	
	Creation	Pakistan
1947	Women gained the right to vote	China
1948	Creation of the Jewish state of Israel	
1949	Universal suffage	India
1950 -	Kofi Annan	
1950	Korean War	
1952	Nasser deposes King Faruk	Egypt
1956	Hungarian uprising against Russia	
	First constitution of Pakistan ratified	
1956	Women gained full voting rights	Pakistan
1961	Yuri Gagarin orbits the earth	Russia
1965	War in Vietnam	
1966	Dr. Maulana Karenga	
	KWANZAA	USA
1967	Six-day war	
1969	Moon Landing	
	Golda Meir premier of Israel	
1970	Invasion of Cambodia	

	Nasser succeeded by Anwar al-Sadat	Egypt
	10,000 women march for equal rights	New York
1973	Egypt and Syria attack Israel	
1978	Camp David Accords	
1979	Ayatollah Khomeini heads Iran	
1981	Anwar al-Sadat assassinated	
	Columbia space shuttle launched	California
1982	Battle of the Falkland Islands	
1983	War in Grenada	
	Space shuttle Challenger arrives at Edwards Air Base	California
1986	Chernobyl accident	Russia
	Challenger explodes	
	Duvalier flees	Haiti
1987	Intifadah	
1990	Saddam Hussein of Iraq invades Kuwait	
1991	Operation Desert Storm	Kuwait
1993	Israel and Palestine sign Oslo Accords	USA
1994	HAMAS suicide bombers begin attacking Israel civilians	
	Yitzak Rabin assassinated	
	Taliban fundamentalists gain control in Afghanistan	
1997	Sayyid Khatami elected president	Iran
2001(9.11)	Destruction of the Twin Towers	New York
	Invasion	Afghanistan
2002	War on Terrorism	Iraq

*　**The War to End All Wars**

If ye break faith with us who die
We shall not sleep
Though poppies grow in Flanders Fields

CANDLELIGHTING

AND

MEDITATION

BEAUTY

Robert Bridges. I love all beauteous things; I seek and adore them. God has no better praise, and humans in their hasty days are honored for them.

John Keats. A thing of beauty is a joy forever.

Kahlil Gibran. Beauty is eternity gazing at itself in a mirror. But you are eternity and you are the mirror.

Gerard Manly Hopkins. All things counter, original, spare, strange; whatever is fickle, freckled (who knows how?) with swift, slow; sweet, sour; dazzle, dim; God fathers-forth whose beauty is past change - Praise God.

Albert Einstein. To know that what is impenetrable to us really exists, manifesting itself as the highest wisdom and the most radiant beauty...this knowledge, this feeling, is at the center of true religiousness.

Socrates. I pray to you, Oh God, that I may be beautiful within.

Harriet Beecher Stowe. In all ranks of life the human heart yearns for the beautiful, and the beautiful things that God makes are a gift to all alike.

Sir Rabindranath Tagore. Beauty and her twin brother Truth require leisure and self-control for their growth.

Anthony Ashley Cooper (Lord Shaftesbury). All beauty is truth. True features make the beauty of the face; true proportions, the beauty of architecture; true measures, the beauty of harmony and music.

Christian Dior. Zest is the secret of all beauty. There is no beauty that is attractive without zest.

Kahlil Gibran. Beauty is not a mouth thirsting nor an empty hand stretched forth, but rather a heart enflamed and a soul enchanted.

Alfred Joyce Kilmer. I think that I shall never see a poem lovely as a tree. Poems are made by fools like me, but only God can make a tree.

Richard Bach.   A cloud does not know why it moves in just such a direction and at such a speed.   But the sky knows the reasons and the patterns behind all clouds and you will know too, when you lift yourself high enough to see beyond horizons.

Christianity (Luke 12:27).   Consider the lilies how they grow; they toil not, neither do they spin; yet I say to you, that Solomon in all his glory was not arrayed like one of these.

Native American (Bear Paw).   Being aware of the many true essences of a thing will give a glimpse into the understanding that all things contain the elements of beauty.

John Keats. Beauty is truth, truth beauty.

Edna St. Vincent Millay.   Lord I fear you made the world too beautiful this year, My soul is almost out of me - let no burning leaf fall, let no bird call.

Mahatma Gandhi.   True beauty consists of purity of heart.

Johann von Goethe.   The soul that sees beauty may sometimes walk alone.

Eleanor Roosevelt.   The future belongs to those who believe in the beauty of their dreams.

Michelangelo.   My soul can find no staircase to heaven unless it be through earth's loveliness.

Ralph Waldo Emerson.   Beauty is God's handwriting - a wayside sacrament. Welcome it in every fair face, in every fair sky, in every fair flower, and thank God for it as a cup of blessing.

Aristotle. Beauty is the gift of God.

Christian Nestell Bovee. To cultivate the sense of the beautiful, is one of the most effective ways of cultivating an appreciation of the divine goodness.

Luther Burbank. The need for beauty is as positive a natural impulse as the need for food.

Richard Cecil. Every year of my life I grow more convinced that it is wisest and best to fix our attention on the beautiful and the good, and dwell as little as possible on the evil and the false.

Emily Dickinson. Beauty is not caused. It is.

Anthony Ashley Cooper. (Lord Shaftesbury). The most natural beauty in the world is honesty and moral truth.

CHILDREN

Elizabeth Barrett Browning. Do you hear the children weeping, O my brothers, ere the sorrow comes with years?

Judith Wright. This is our hunter and our chase, the third who lay in our embrace.

Sappho. I have a small daughter called Cleis. I wouldn't take all Croesus' kingdom with love thrown in, for her.

William Shakespeare. O lord!, my fair son! My life, my joy, my food, my all in the world! My widow-comfort, and my sorrow's cure.

Dennis J. Kucinich. Our children deserve a world free of the terror of hunger, free of the terror of poor health care, free of the terror of homelessness, free of the terror of ignorance, free of the terror of hopelessness.

UNICEF (Farida Ali). Every aspect of a child's life is a human rights issue.

Joseph Joubert. Children have more need of models than of critics.

J. Robert Oppenheimer. There are children playing in the street who could solve some of my top problems in physics, because they have modes of sensory perception that I lost long ago.

Pablo Casals. You must cherish one another. You must work - we all must work - to make this world worthy of its children.

Henry Wadsworth Longfellow. A torn jacket is soon mended, but hard words bruise the heart of a child.

Alexander Pope. Just as the twig is bent, the tree is inclined.

Elizabeth Barrett Browning: But the young children are weeping in the playtime of the others, in the country of the free.

Charles Dickens. I love these little people; and it is not a slight thing, when they, who are so fresh from God, love us.

Cicero. What gift has Providence bestowed on humans that is so dear to them as their children?

George Eliot (Mary Ann Evans). In the person whose childhood has known caresses and kindness, there is always a fibre of memory that can be touched by gentle issues.

John Milton. Childhood shows the adult, as morning shows the day.

Mary Botham Howitt. God sends children to enlarge our hearts; and to make us unselfish and full of kindly sympathies and affections; to give our souls higher aims; to call out all our faculties to extended enterprise and exertion.

James Russell Lowell. Children are God's apostles, sent forth, day by day, to preach of love and hope and peace.

Adam Gottlieb Ochlenshlager. The plays of natural lively children are the infancy of art. Children live in a world of imagination and feeling. They invest the most insignificant object with any form they please, and see in it whatever they wish to see.

Jean Paul Richter. The smallest children are nearest to God, as the smallest planets are nearest the sun.

Oscar Wilde. The best way to make children good is to make them happy.

William Wordsworth. The child is father of the adult.

Charles Buston. The first duty to children is to make them happy. If you have not made them so, you have wronged them. No other good they may get can make up for that.

Samuel Taylor Coleridge. I have often thought what a melancholy world this would be without children; and what an inhuman world, without the aged.

Anne Bradstreet. Thus gone, amongst you I may live, and dead, yet speak, and counsel give: farewell, my birds; farewell, adieu, I happy am, if well with you.

John Dewey. What the best and wisest parents want for their own child, that must the community want for all its children.

Baron Friedrich von Hardenberg (Novalis). Where children are, there is the golden age.

Robert Henri. Feel the dignity of a child. Do not feel superior, for you are not.

Charles and Mary Lamb. You straggler into loving arms, young climber-up of knees, when I forget your thousand ways, then life and all shall cease.

Kahlil Gibran. Your children are not your children. They are the sons and daughters of Life's longing for itself.

COMMUNITY

Albert Einstein. Individuals are what they are and have the significance that they are, not so much by virtue of their individuality, but rather as a member of a great human community, which directs their material and spiritual existence from the cradle to the grave.

David W. Orr.   By community I mean, rather, places in which the bonds between people and those between people and the natural world, create a pattern of connectedness, responsibility and mutual need.

Group Dialogue at Eco Village.   Community is the attractive power or cohesive force of shared reality.

Henrik Ibsen.   A community is like a ship; everyone ought to be prepared to take the helm.

William James.   The community stagnates without the impulse of the individual. The impulse dies away without the sympathy of the community.

Richard Moe. Communities can be shaped by choice or ... by chance. We can keep accepting the kind of communities we get, or we can insist on getting the kind of communities we want.

Group Dialogue at Eco Village.   Community is a human culture formed when the factors which cause human beings to bond are stronger than the factors which tend to divide them.

William Somerset Maugham.   Conscience is the guardian in the individual of the rules which community has evolved for its own preservation..

Charles Kuralt. Communities are all different, but I know when I'm in one. People speak to their neighbors by name, sometimes also to their neighbor's dogs. No one is a stranger for very long.

David W. Orr.   Real communities foster dignity, competence, participation and opportunities for good work. And good communities provide places in which children's imagination and earthly sensibilities root and grow.

John Ruskin.   When we build, let us think that we will build forever.

Lewis Mumford.   Above all we need, particularly as children, the reassuring presence of a visible community, an intimate group that enfolds us with understanding and love, and that becomes an object of our spontaneous loyalty, as a criterion and point of reference for the rest of the human race.

John W. Gardner.   In a system in which responsibility is not concentrated at the center, everyone had better be partly responsible. So it is for a community, so it is for a nation - and, most poignantly, so it is for the only liveable planet we know about.

Edmund N. Bacon.   True involvement comes when the community and the designer turn the process of planning the city into a work of art.

Ralph Nader. When strangers start acting like neighbors, communities are reinvigorated.

EQUALITY OF MEN AND WOMEN

Baha'i. Both men and women are equal in the sight of God, and both have the same rights and responsibilities.

Desmond (Archbishop) Tutu.   There can be no true liberation that ignores the liberation of women.

Hans Selve.   A long, healthy, and happy life is the result of making contributions, of having meaningful projects that are personally exciting and contribute to and bless the lives of others.

Jack Kornfield.   Live every act fully, as if it were your last. No matter how difficult the past, you can always begin again today.

George MacDonald.   If I can put one touch of a rose sunset into the life of any man or woman, I shall feel that I have worked with God.

Christianity (James 1:4).   Let patience have our perfect work that we may be perfect and entire, wanting nothing.

Native American (Bear Paw).   Women and men exist under a humanness that isn't measured with equality. It is only defined in the sole sense of being human. We are as we are, all human. All other physical and mental abilities are only measurements created by a social environment. Both sexes are inseparable - one cannot exist without the other; both are one.

Jack Kornfield. At the bottom of things, most people want to be understood and appreciated.

Baha'i (Abdu'l-Baha). The world of humanity has two wings - one is feminine and the other masculine. Not until both wings are equally developed can the bird fly.

Zorastrianism. All men and women are members of one human species - one family.

Anna Wickham. Not equal I, but counterpart and in relation is my heart; perfect with man's - as with his mind - mine is all strong to loose and bind.

James Dillet Freeman. May you always need one another - not so much to fill your emptiness as to help you to know your fullness.

Mary Wollstonecraft. If women are by nature inferior to men, their virtues must be the same in quality, if not in degree, or virtue is a relative idea; consequently, their conduct should be founded on the same principles and have the same aim.

ETHICS

Junius. The integrity of people is to be measured by their conduct, not by their professions.

Mahatma Gandhi. I do feel that there is orderliness in the universe, that there is an unalterable law governing everything and every being that lives and moves. The law and the lawgiver are one.

Cardinal John Henry Newman. All virtue and goodness tend to make me powerful in the world; but those who aim at the power have not the virtue.

Sophocles. To be doing good is human's most glorious task.

Carol Gilligan. While women thus try to change the rules in order to preserve relationships, men, in abiding by these rules, depict relationships as easily replaced.

Cyril Edwin Mitchinson Joad. Goodness must be freely acquired; it cannot be imposed from without by discipline, and cannot be achieved by merely keeping the rules.

Ken Carey. The choice of love may not always seem like the swiftest path to resolution of difference, but it is the only certain one.

Joe Delagarza. You are meeting yourself daily, and everyone you meet is a representation of you and your karma.

Albert Schweitzer. Every profound philosophy, every deep religion is a search for ethical mysticism and mystical ethics.

Epictetus. In all the affairs of life let it be your great care not to hurt your mind, or offend your judgment.

Baha'i (Abdu'l-Baha). Old trees yield no fruitage; old ideas and methods are obsolete and worthless now. Old standards of ethics, moral codes and methods of living in the past will not suffice for the present age of advancement and progress.

Wayne Dyer. In judging others, we define ourselves.

Aristotle. Humans perfected by society are the best of all animals; they are the most terrible of all when they live without law and without justice.

General Douglas MacArthur. I have known war as few people living know it. Its very destructiveness on both friend and foe has rendered it useless as a means of settling international disputes.

Zoroastrianism. Life is best lived when it is lived for others.

Frederick Dan Huntington. What people do, tells us who they are.

Independent. There may be a certain pleasure in vice, but there is a higher reward in purity and virtue. It is the loss of the sense of sin and shame that destroys both humans and states.

Thomas Henry Huxley.   My belief is that no human beings or societies ever did, or ever will, come to much unless their conduct was governed and guided by the love of some ethical idea.

Christianity (Pius XII).   Though economic science and moral discipline are guided each by its own principles, it is false that the two orders are so distinct that the former in no way depends upon the latter.

Thomas Jefferson. A lively and lasting sense of filial duty is more effectually impressed on the mind by reading *King Lear*, than by all the dry volumes of ethics and divinity.

William James.   The ultimate test for us of what a truth means is the conduct it dictates or inspires.

Richard Kimball. It is unethical to ask someone to lie to advance personal or institutional objectives.

FAITH

Asa Gray .   Faith in order, which is the basis of science, cannot reasonably be separated from faith in an Ordainer, which is the basis of religion.

Mahatma Gandhi.   Though, from my weakness, I fail a thousand times, I will not lose faith.

Judaism (Hebrews 11:1).   Faith is the assurance of things hoped for and the conviction of things not seen.

Pearl S. Buck. All things are possible until they are proved impossible - and even the impossible may only be so as of now.

Baha'i. By faith is meant first, conscious knowledge, and second, the practice of good deeds.

A Course in Miracles. Nothing outside yourself can save you - nothing outside yourself can give you peace - nothing outside yourself can hurt you. Nothing but your own thoughts can hamper your progress. You are free from all external interference.

Mahatma Gandhi. The truth is that God is the force. It is the essence of life. It is pure, undefiled consciousness. It is eternal.

Native American (Bear Paw). To feel the ground below our feet is to feel a confidence and trust in life.

Christianity (Philippians 1:6) He that has begun a good work in you will perform it until the day of completion.

Stella Terrill Mann. Desire, ask, believe, receive.

Anonymous. All the darkness of the world cannot put out the light of one small candle.

Emily Bronte. No coward soul is mine, no trembler in the world's storm-troubled sphere; I see Heaven's glories shine, and faith shines equal, arming me from fear.

Frances Beaumont. Faith without works is like a bird without wings; but when both are joined together, then the soul mounts up to her eternal rest.

Louis Bromfield. Human beings are not naturally cynics; they want to believe in themselves, in their future, in their community and in the nation of which they are a part.

Frank Crane. You may be deceived if you trust too much, but you will live in torment if you do not trust enough.

Mormon. Power is given to do all things by faith.

Christianity (Matthew 6:34). So do not be anxious about tomorrow; tomorrow will look after itself. Each day has troubles enough of its own.

Jane Adams. What after all has maintained the human race on this old globe, despite all the calamities of nature and all the tragic failings of humankind, if not the faith in new possibilities and the courage to advocate them?

FAMILY

Felix Adler. The family is the miniature commonwealth, upon whose integrity the safety of the larger commonwealth depends.

William Aikman. Civilization varies with the family, and the family with civilization.

Sir John Bowring. A happy family is but an earlier heaven.

Horace Bushnell. A house without a roof would scarcely be a more different home than a family unsheltered by God's friendship.

Robert Lee Frost. The greatest thing in family life is to take a hint when a hint is intended - and not to take a hint when a hint isn't intended.

Robert Francis Kennedy. In my judgment, one of the basic reasons we have crime, lawlessness, and disorder has been the breakdown of the family unit.

Jean Francois Marmontel. Where can individuals better be than with their families?

William Ellery Channing. The ties of family and of country were never intended to circumscribe the soul. If allowed to become exclusive, engrossing, clannish, so as to shut out the general claims of the human race, the highest end of Providence is frustrated. and the home, instead of being the nursery, becomes the grave of the heart.

Leo XIII (Gioacchino Vincenzo Pecci). The family is a society limited in numbers, but nevertheless a true society, anterior to every state or nation, with rights and duties of its own wholly independent of the commonwealth.

William Makepeace Thayer. As are families, so is society. If well ordered, well instructed, and well governed, they are the springs from which go forth the streams of national greatness and prosperity - of civil order and public happiness.

Sioux Indian. With all beings and all things we shall be as relatives.

Morihei Ueshiba. All things, material and spiritual, originate from one source and are related as if they were one family.

FORGIVENESS

Desmond Tutu. Forgiveness is the grace by which you enable the other person to get up, and get up with dignity, to begin anew.

Confucius. Those who cannot forgive others break the bridge over which they themselves must pass.

Judaism. (Proverbs 19:11). The discretion of people defers anger; and it is their glory to pass over a transgression.

Anonymous. No matter how good your friends are, they are going to hurt you every once in a while, and you must forgive them for that.

Christianity. (Luke 6:37). Forgive, and you shall be forgiven.

William Shakespeare. To err is human, to forgive divine.

Alice Bailey. Forgiveness and sacrifice are increasingly needed keynotes at this time.

Norman Cousins. Life is an adventure in forgiveness.

Judaism. (Isaiah 1:18). Though your sins be as scarlet, they shall be as white as snow; though they be red like crimson, they shall be as wool.

Richard Bach. You are led through your lifetime by the inner learning creature, the playful spiritual being that is your real self.

Jainism. Forgive all creatures and let all creatures forgive me.

Christianity (John 1:9). If we confess our sins, Spirit is faithful and just to forgive us our sins, and to cleanse us from all unrighteousness.

Jessamyn West. It is very easy to forgive others their mistakes; it takes more grit and gumption to forgive them for having witnessed your own.

A Course in Miracles. Forgiveness is my function as the light of the world. My function, my salvation and my happiness are one and the same .

Hindu poet (19ᵗʰ Century). They forgot all the good turns we did for them. When they came back, they gave us so much pain. Oh, how long should we take to forgive them? Do not wait, do it today, forgive them.

Christianity. (Luke 23:34). Father forgive them, for they know not what they do.

Shirley Leighton. To forgive is the only way to establish equanimity within the confines of oneself.

Jack Kornfield. Most of the sorrows of the earth humans cause for themselves.

Anonymous. It isn't enough to be forgiven by others. Sometimes we have to learn to forgive ourselves.

Hannah More. Forgiveness saves the expense of anger, the cost of hatred, the waste of spirits.

Madame de Stael. The more we know, the better we forgive. Whoever feels deeply, feels for all that live.

Christianity (Ephesians 4:31). Put away from you all bitterness and wrath...and be kind to one another, tenderhearted, forgiving one another.

William Shakespeare. I pardon you, as heaven shall pardon me.

FULFILMENT

Self-Realization Fellowship (Paramahansa Yogananda). Understanding is the searchlight that illumines your way and brings success.

Ray Bradbury. We are cups, constantly and quietly being filled. The trick is to know how to tip ourselves over and let the beautiful stuff out.

Bawa Muhaiyaddeen. Place patience in front of you. Give all praise to God, and then go forward. That will be your plenitude.

Jack Kornfield. Live every act fully, as if it were your last.

Christianity (Jesus). Do not be anxious about tomorrow, for tomorrow will take care of itself. Let the day's challenges be sufficient for the day.

Christian D. Larson. A genius is asleep in the subconscious of every mind; a spiritual giant is within us awaiting recognition.

Judaism (Psalms 55:22). Cast your burden on the Lord, and God shall sustain you.

A. D. Hope. There was a time the poet's mission was to give humans their daily bread, the crown of life, the timeless vision which linked the living with the dead.

Christianity (Matthew 16:25). Those who will lose their lives for My sake shall find them.

Hazrat Inayat Khan. The words that enlighten the soul are more precious than jewels.

Glenda Green. So when you pray, do so in joy for that which has been promised: pray for guidance that your mind may work in harmony with your heart, to direct your life toward its fulfillment.

Native American (Bear Paw). Without any apprehension, be true to your inner needs, dreams and thoughts and you'll find that you are entering into a state of encompassing all that is in harmony with a living cell.

Henry David Thoreau. Dwell as near as possible to the channel in which your life flows.

Jonas Edward Salk. I feel that the greatest reward for doing is the opportunity to do more.

Helen Adams Keller. We can do anything we want to do if we stick to it long enough.

Arthur Wilberforce Jose. This is the sum of things: - that we in a lifetime live great-heartedly; see the whole best that life has meant; do our work, and go content.

Christina Georgina Rossetti. Shall I find comfort, travel-sore and weak? Of labor you shall find the sum. Will there be beds for me and all who seek? Yes, beds for all who come.

Apocrypha (Ecclesiasticus, Ch. xliv) Their seed shall remain for ever, and their glory shall not be blotted out; their bodies are buried in peace; but their name lives for evermore.

Andrew Barton (Banjo) Patterson. They see the vision splendid of the sunlit plains extended, and at night the wondrous glory of the everlasting stars.

GOLDEN RULE

Christianity. All things which you want people to do to you, do also to them.

Baha'i (Abdu'l-Baha). The first remedy is to guide the people, and go forth with a hearing ear and seeing eye.

Confucianism. Do not do to others what you would not want them to do to you.

Bawa Muhaiyadden. Do not harm another even in your thoughts, for it will result in greater harm to you.

Buddhism. Clansmen minister to their friends by treating them as they treat themselves.

Immanuel Kant. Act in conformity with that maxim which you can at the same time will to be a universal law.

Hinduism. Do not do to others what if done to you, would cause you pain.

Islam. No one of you is a believer until you love for others what you love for yourself.

Albert Schweitzer. Ethics consist in my experiencing the compulsion to show all will-to-live the same reverence as I do to my own life.

Sikhism. As you value yourself, so value others. Then shall you become a partner in the spirit.

Judaism. What is hurtful to yourself, do not do to your fellow beings.

William Shakespeare. To your own self be true, and it must follow as the night the day, you cannot then be false to anyone.

Jainism. In happiness and suffering, in joy and grief, we should regard all creatures as we regard our own self.

Zorastrianism. That nature only is good when it shall not do to another whatever is not good for its own self.

Bawa Muhaiyaddeen. Do not analyze the right and wrong in another: analyze and understand the right and wrong within your own life. That will be of greater benefit to you.

Taoism. Regard your neighbor's gain as your own gain and regard your neighbor's loss as your own loss.

Martin Luther King, Jr. Sooner or later all the peoples of the world will have to discover a way to live together.

Judaism. You shall love the Lord your God with all your heart and your neighbor as yourself.

Zorastrianism. Happiness belongs to those who promote the happiness of others. Teach us to seek our happiness in the happiness of all, to regard the sorrows and suffering of others as ours and hasten to assuage them.

GOVERNMENT

Eleanor Roosevelt. I believe we will have better government when men and women discuss public issues together and make their decisions on the basis of their common concern for the welfare of their families and their world..

William Shakespeare. For government, through high, and low and lower, put into parts, does keep in one consent; congreeing in a full and natural close, like music.

Mahatma Gandhi. A government is an instrument of service only in so far as it is based upon the will and consent of the people. It is an instrument of oppression when it enforces submission at the point of the bayonet. Oppression therefore ceases when people cease to fear the bayonet.

Johann Wolfgang von Goethe. The best of all governments is that which teaches us to govern ourselves.

Plato. The punishment suffered by the wise who refuse to take part in the government, is to live under the government of bad people.

Dwight David Eisenhower. We instinctively have greater faith in the counterbalancing effect of many social, philosophic and economic forces than in arbitrary law.

Ralph Waldo Emerson. The less government we have the better - the fewer laws and less confided power. The antidote to this abuse of formal government is the influence of private character, the growth of the individual.

Ralph Tyler Flewelling. It is a perversion to call that a state of freedom where one class rules to the disadvantage of another. A benevolent despotism may be more tolerable than an intolerant class democracy.

Mahatma Gandhi. Good government is no substitute for self-government.

William Gladstone. The proper function of a government is to make it easy for the people to do good and difficult for them to do evil.

Sir James Hopwood Jeans. The actual achievement of democracy is that it gives a tolerably good time to the underdog. Or, at least, it tries, and it is, I think, for this reason that most of us accept it as our political creed.

John Fitzgerald Kennedy. Government is an art and a precious obligation; and when it has a job to do, I believe it should do it. And this requires not only great ends but that we propose concrete means of achieving them.

Herbert Henry Lehman. You can't run a government solely on a business basis. Government should be human. It should have a heart.

Abraham Lincoln. This nation, under God, shall have a new birth of freedom, that government of the people, by the people, for the people, shall not perish from the earth.

Donald Robert Perry (Don) Marquis. The bees got their governmental system settled millions of years ago but the human race is still groping.

Philip Massinger. He that would govern others, first should be the master of himself, richly endued with depth of understanding and height of knowledge.

Lewis Mumford. Before we can have a sound village government, we must have a world government. Families cannot be permanently united with any prospect of a good life together until humankind is united.

Jawaharlal Nehru. A government in a democratic society is a reflection of the will of the people, and it should continue to be a reflection of this all the time.

Ralph Barton Perry. Any government has in the last analysis to be justified by the quality of the life which it promotes.

Sir Walter Raleigh. People well governed should seek after no other liberty, for there can be no greater liberty than a good government.

Franklin Delano Roosevelt. Better the occasional fault of a government that lives in a spirit of charity than the consistent omissions of a government in the ice of its own indifference.

145

Charles de Secondat (Baron de Montesquieu). The deterioration of every government begins almost always by the decay of the principles on which it was founded.

Sri Aurobindo. If we are intended to survive and carry forward the evolution of which humanity is at present the head, we must come out of our present chaotic international life and arrive at a beginning of organized united action.

Solon. That is the most perfect government under which a wrong to the humblest is an affront to all.

Baruch (Benedict) Spinoza. The ultimate aim of government is.... to free every person from fear, that each one may live in all possible security....In fact, the true aim of government is liberty.

Adlai Ewing Stevenson. It seems to me that government is like a pump, and what it pumps up is just what we are, a fair sample of the intellect, the ethics and the morals of the people, no better, no worse.

George Washington. While just government protects all in their religious rites, true religion affords government its surest support.

(Thomas) Woodrow Wilson. Only free people can hold their purpose and their honor steady to a common end, and prefer the interest of humankind to any narrow interest of their own.

GRATITUDE

Anne Sophie Swetchine. Those who make us happy are always thankful to us for being so; their gratitude is the reward of their benefits.

Seneca. If I only have the will to be grateful, I am so.

Pierre Jean de Beranger. Those who remember the benefits of their parents are too much occupied with their recollections to remember their faults.

Richard Kimball. Offering our skills, services, or time for listening can be a true expression of our gratitude and appreciation.

Rudolph Block (Bruno Lessing). A grateful thought toward heaven is of itself a prayer.

Thomas Secker. A grateful mind is both a great and a happy mind.

Alexander Pope. When I find a great deal of gratitude in poor people, I take it for granted there would be as much generosity if they were rich.

Benjamin Franklin. To the generous mind the heaviest debt is that of gratitude when it is not in our power to repay it.

William Shakespeare. I thank you for your voices, - thank you, - your most sweet voices.

Felix Frankfurter. Gratitude is one of the least articulate of the emotions, especially when it is deep.

William Romaine. Gratitude to God makes even a temporal blessing a taste of heaven.

Pierre Charron. A person who receives a benefit should never forget it; a person who bestows should never remember it.

Seneca. There is as much greatness of mind in acknowledging a good turn, as in doing it.

Charles Simmons Gratitude to God should be as habitual as the reception of mercies is constant.

Jeremy Taylor. Learn to give thanks for everything.

Nathaniel Parker Willis. Gratitude is not only the memory but the homage of the heart - rendered to God for all goodness.

Shirley Leighton. Forgiveness is love, it is peace, it is understanding. Most of all, it is humble gratitude.

Richard Kimball. We can also be generous when we receive, for when we receive, we allow other persons to give and we support their establishing a bond with us that may not have existed before.

William Shakespeare. I can no other answer make, but thanks, and thanks: and ever often good turns are shuffled off with such uncurrent pay.

HARMONY

St. Theresa of Lisieux. Each small task of everyday life is part of the total harmony of the universe.

Jack Kornfield. In one's family, respect and listening are a source of harmony.

Albert Einstein. My gifts of fantasy are more powerful than my ability to assume more knowledge.

Indian Wisdom. Listen, or your tongue will keep you deaf.

Morihei Ueshiba. The universe emerged and developed from one source, and we evolved through the optimal process of unification and harmonization.

Peter Marshall. When we long for life without difficulties, remind us that oaks grow strong in contrary winds and diamonds are made under pressure.

Marc Chagall. Our whole inner world is reality - perhaps more real than the apparent one.

William Shakespeare. Look how the floor of heaven is thick inlaid with patines of bright gold. Such harmony is in immortal souls, but while this muddy vesture of decay closes it in, we cannot hear it.

Cadillac. Though we should soar into the heavens, though we should sink into the abyss, we never go out of ourselves; it is always our own thought that we perceive.

Alexander Pope. Here earth and water seem to strive again, not chaos-like, but as the world, harmoniously confused, where we see order in variety, and where, though all things differ, all agree.

Jack Kornfield. Live like the strings of a fine instrument - not too taut and not too loose.

Science of Mind (Ernest Holmes). The secret is getting into a right relationship with the Universal. In tune with Infinite Order is harmony. Such harmony adjusts our affairs and enriches our lives.

Thich Nhat Hanh. If in our daily life, we can smile, not only we, but everyone will profit from it.

Vedanta (Swami Vivekananda). Education is the manifestation of the perfection already in us.

Shirley Leighton. When illusions cause pain, only stillness informs us of the reality where love and peace abide.

Taoism (Lao Tze). Knowing others is wisdom. Knowing oneself is enlightenment.

Douglas Jerrold. Humor is the harmony of the heart.

William James. The art of being wise is the art of knowing what to overlook.

John Dryden. From harmony, from heavenly harmony, this universal frame began.

Elizabeth Barrett Browning. And cottage-gardens smelling everywhere, infused with smell of orchards. "See," I said, "And see! Is God not with us on the earth?"

W.H. Murray. In the architecture of hill and sky, as in great art and music, there is an everlasting harmony with our own being.

HEALTH AND HEALING

Science of Mind. Lifelong habits of wrong thinking can be consciously and deliberately neutralized.

Norman Vincent Peale. Nobody can be you as efficiently as you can.

149

Christianity (Luke 12:31). Seek first the kingdom of God, and all these things shall be given to you.

Shirley Leighton. Depression, anxiety or other forms of illness are relieved by forgiveness.

Deepak Chopra. Impulses of intelligence constantly create the body in new forms every second.

Shakti Gawain. An affirmation is a strong positive statement that something already is.

J. Krishnamurti. Real learning comes about when the competitive spirit has ceased.

Christianity (Luke 11:9). And I say to you, ask and it shall be given you; seek and you shall find; knock, and it shall be opened to you.

Indian Wisdom. Listen or your tongue will keep you deaf.

Carl Gustav Jung. Explore daily the will of God.

Mother Teresa. Find people who think they are alone and let them know that they are not.

Eckhart Tolle. When you are suffering, when you are unhappy, stay totally with what is Now. Unhappiness or problems cannot survive in the Now.

Jack Kornfield. Blessings come from care, trouble from carelessness.

Zoroastrianism. Our prime requirement is to be happy with a sound body and a clear brain.

Anonymous. If you're not evolving, you're decaying.

Eckhart Tolle. Naming something as bad causes an emotional contraction within you. When you let it be, without naming it, enormous power is suddenly available to you.

Arabian Proverb. Those who have health, have hope; and those who have hope, have everything.

Mahatma Gandhi. It is health which is real wealth and not pieces of gold and silver.

HOPE

Dalai Lama. The hope of all people, in the last analysis, is simply for peace of mind.

Samuel Johnson. Where there is no hope, there can be no endeavor.

John Greenleaf Whittier. The dark night is over and dawn has begun. Rise, hope of the ages, rise like the sun! All speech flows to music, all hearts beat as one.

World Peace Prayer. Lead me from death to life - from falsehood to truth. Lead me from despair to hope - from fear to trust. Lead me from hate to love - from war to peace. Let peace fill our hearts, our world, our universe.

Zoroastrianism. The soul of a person may be daring and eager to work for righteousness and to fight wickedness within and without.

Martin Luther King, Jr. We shall hew out of the mountain of despair, a song of hope.

Bruce Barton. Before you give up hope, turn back and read the attacks that were made upon Abraham Lincoln.

Johann Wolfgang von Goethe. In all things it is better to hope than to despair.

David Hume. A propensity to hope and joy is real riches; one to fear and sorrow, real poverty.

Samuel Johnson. Whatever enlarges hope will also exalt courage.

Robert Francis Kennedy. Each time people stand for an ideal, or act to improve the lot of others, or strike out against injustice, they send forth a tiny ripple of hope.

Thales. Hope is the only good that is common to all people; those who have nothing else possess hope still.

John La Farge. The prophet of despair gains a shouting audience, but one who speaks from hope will be heard long after the noise dies down.

Alexander Pope. Hope springs eternal in the human breast; humanity never is but always to be blest.

Johann Wolfgang von Goethe. You resemble the thought which you conceive.

William Bolitho. The adventurer is within us and contests for our favor with the social people we are obliged to be.

Michael Guillen. Dimension is in the eye of the beholder.

HUMAN RIGHTS AND RESPONSIBILITIES

John Fitzgerald Kennedy. In giving rights to others which belong to them, we give rights to ourselves and to our country.

Nelson Mandela. Let the strivings of us all prove Martin Luther King Jr. to have been correct when he said that humanity can no longer be tragically bound to the starless midnight of racism and war.

Baha'i (Abdu'l-Bah'a). This is a brilliant century. Eyes are now open to the beauty of the oneness of humanity. The darkness of suppression will disappear, and the light of unity will shine.

Richard Bach. Every person and all the events of your life are there because you have drawn them there. What you choose to do with them is up to you.

Elizabeth Charlotte Pauline de Guizot. Much misconstruction and bitterness are spared to those who think naturally upon what they owe to others rather than what they ought to expect from them.

Native American(Bear Paw). Human rights are based on the essence of the contribution of the individual to the whole. All races should have rights that focus on their own unique differences. No race has permission to create rights for a race other than itself.

George Bancroft. The fears of one class of people are not the measure of the rights of another.

Anonymous. Our background and circumstances may have influenced who we are, but we are responsible for who we become.

Nelson Mandela. What challenges us is to ensure that none should enjoy lesser rights, and none tormented because they are born different, hold contrary political views, or pray to God in a different manner.

Jack Kornfield. No one can help us as much as our own compassionate thought.

Mahatma Gandhi. In fact, the right to perform one's duties is the only right that is worth living for and dying for. It covers all legitimate rights.

Carl Schurz. From equality of rights springs identity of our highest interests; you cannot subvert your neighbor's rights without striking a dangerous blow at your own.

Harry S. Truman. In the cause of freedom, we have to battle for the rights of people with whom we do not agree. If we do not defend their rights, we endanger our own.

Francis Lieber. There is no right without a parallel duty, no liberty without the supremacy of the law, no high destiny without earnest perseverance, no greatness without self-denial.

IMMORTALITY

Socrates. If it be true that the soul is immortal, we have to take care of her, not merely on account of the time which we call life, but also on account of all time.

Ella Wheeler Wilcox. The spark divine dwells in our souls, and we can fan it to a steady flame of light, whose luster gilds the pathway to the tomb and shines on through Eternity.

Judaism. The dust returns to the earth as it was, and the Spirit returns to God who gave it.

Christianity. The gift of God is eternal life.

Science of Mind (Ernest Holmes). We are not going to attain immortality; we are now immortal. It is not that the dead live again, but that the living never die.

Langston Hughes. Don't you fall now for I's still goin', honey, I's still climbin', and life for me ain't been no crystal stair.

Jainism. I know there will be a life hereafter.

Baha'i (Abdu'l-Baha). Let all your striving be for this, to become the source of life and immortality, and peace and comfort and joy, to every human soul.

The Apocrypha. Though they be punished in the sight of mortals, yet is their hope full of immortality.

Confucianism. All the living must die and, dying, return to the ground, but the Spirit issues forth and is displayed in light.

Kathleen Raine. I am light traveling in empty space. I am a diminishing star speeding away out of the universe.

James Shirley. Only the actions of the just smell sweet and blossom in their dust.

Hinduism.　The person becomes immortal who seeks the general good of humanity.

Martin Luther King, Jr.　The sacred heritage of our nation and the eternal Will of God are embodied in our echoing demands.

Sikhism.　Why weep when a person dies, since that one is only going home.

Sappho.　Although they are only breath, words which I command are immortal.

Buddhism.　Earnestness is the path of immortality.

Emily Dickinson.　Because I could not stop for death, death kindly stopped for me; the carriage held but just ourselves and immortality.

Shinto.　Regard Heaven as your father, earth as your mother, all things as brothers and sisters, and you will enjoy the divine country which excels all others.

Marlo Thomas.　All humans are spirits only visiting this world. All spirits are forever beings. All encounters with other people are experiences and all experiences are forever connections.

Taoism.　Life is going forth. Death is a returning home.

Elizabeth Barrett Browning.　I love you with the breath, smiles, tears of all my life! - and, if God choose, I shall but love you better after death.

Zoroastrianism.　The soul of the righteous shall be joyful in immortality.

Vedanta (Swami Vivekananda).　Every evolution is preceded by an involution. The seed is the father of the tree, but another tree was itself the father of the seed.

Albert Einstein.　The most beautiful thing we can experience is the mysterious.

Grace Perry. For me also immortality is at my feet, children already overshadow me.

Desmond Tutu. Through our connection to all life and to God, we are stronger than death.

INNER VISION

Judaism (Jeremiah 29:13). And you shall seek me and find me when you search for me with all your heart.

Science of Mind (Ernest Holmes). Like the creative soil in which seeds are planted, the soul of the universe is the creative medium into which the word of Spirit falls and from which creation rises.

Harry S. Truman. Every good thing in the world was once only a dream. Dream your own great dream.

William Blake. See a world in a grain of sand and a heaven in a wildflower, hold infinity in the palm of your hand and eternity in an hour.

Black Elk. Some little root of the sacred tree still lives. Nourish it, that it may leaf and bloom and fill with singing birds.

Baha'i (Abdu'l-Baha). We we must live with heart and soul in order that the veil covering the eye of inner vision may be removed.

John Muir. Everybody needs beauty as well as bread, places to play in and pray in, where nature may heal and cheer, and give strength to body and soul.

Mahatma Gandhi. God is an unseen power residing within us.

Albert Einstein. Imagination is...the preview of life's forthcoming attractions.

Marc Chagall. Our whole inner world is reality, perhaps more real than the apparent one.

Judaism (Proverbs 23:7). For as people think in their hearts, so are they.

Louise Imogen Guiney. We leap to the infinite dark like sparks from the anvil. You lead, O God! All's well with your troopers that follow.

Christianity (Matthew 6:21). Where your treasure is, there will your heart be also.

Leo Tolstoy. True life is lived when tiny changes occur.

Abraham Maslow. A musician must make music, an artist must paint, poets must write if they are to be at peace with themselves. What we can be, we must be.

JOY AND SORROW

Mother Teresa. Joy is a net of love by which you catch souls.

Mormon. How beautiful shall be those who publish tidings of joy.

Burton Hills. Happiness is not a destination. It is a method of life.

Fortune City (web site). It is only when we love someone that we suffer when they do...sadly, grief is often the price we pay for love.

Jack Kornfield. Joy and openness come from our contented heart.

Christianity (Psalms 100. 1,2). Make a joyful noise to the Lord. Serve the Lord with gladness. Come before God's presence with singing.

Native American (Bear Paw). Hope is the spirit of the imagination.

W. B. Ullathorne. Nothing contributes more to cheerfulness than the habit of looking at the good side of things.

Agnes Repplier. To withhold from children some knowledge - apportioned to their understanding - of the world's sorrows and wrongs is to cheat them of their kinship with humanity.

Jack Kornfield.   Every life has a measure of sorrow. Sometimes it is this that awakens us.

Zoroastrianism. We should live well and enjoy life's bountiful gifts, though in moderation.

Victor Borge. Laughter is the shortest distance between two people.

Elsa Maxwell. Laugh at yourself first - before anyone else can.

Harvey Mindess. Our sense of humor ranges beyond jokes, beyond wit, beyond laughter itself. It is a frame of mind, a healthy manner of perceiving and experiencing life.

Fred Allen. Our childish delight in nonsense must be given free reign again, for humor responds to absurdity like a flower to the sun.

H.L. Mencken.   It is a dull person who is always sure, and the sure person who is always dull.

Henry Ward Beecher.   Joy is more divine than sorrow, for joy is bread and sorrow is medicine.

Samuel Langhorne Clemens (Mark Twain).   Grief can take care of itself, but to get full value of a joy you must have somebody to divide it with.

Helen Adams Keller. The world is so full of care and sorrow that it is a gracious debt we owe to one another to discover the bright crystals of delight hidden in somber circumstances.

Paul Kawken. Always leave enough time in your life to do something that makes you happy, satisfied, even joyous. That has more of an effect on your well-being than any other single thing.

JUSTICE

Aristotle. Justice is to give to all people their own.

Dwight David Eisenhower. Though force can protect in emergency, only justice, fairness, consideration and cooperation can finally lead people to the dawn of eternal peace.

Benjamin Franklin. Justice is as strictly due between neighbor nations, as between neighbor citizens.

Socrates. What is in conformity with justice should also be in conformity to the laws.

Yiddish Proverb. Rather suffer an injustice than commit one.

Voltaire. The sentiment of justice is so natural, and so universally acquired by all humankind, that it seems to be independent of all law, all party, all religion.

Daniel Webster. Justice is the great interest of humans on earth. It is the ligament which holds civilized beings and civilized nations together.

John Jacques Rousseau. An honest person nearly always thinks justly.

Arthur Hendrick Vandenberg. Expediency and justice frequently are not even on speaking terms.

Joseph Addison. Justice discards party, friendship and kindred, and is therefore represented as blind.

William Ewart Gladstone. Justice delayed is justice denied.

Denis Diderot. Justice is the first virtue of those who command and stops the complaints of those who obey.

Blaise Pascal. Justice and power must be brought together, so that whatever is just may be powerful, and whatever is powerful may be just.

John Fitzgerald Kennedy. The achievement of justice is an endless process.

Judaism (Amos 5:24). Justice without wisdom is impossible. But let justice run down as waters, and righteousness as a mighty stream.

Martin Luther King, Jr. It is not enough for us to talk about love; love is one of the pivotal points of the Christian faith. There is another side called justice. And justice is really love in calculation. Justice is love correcting that which revolts against love.

Pope Paul VI. Nothing is lost by peace. Everything may be lost by war.

Harriet Elizabeth Beecher Stowe. In the gates of eternity the black hand and the white hold each other with an equal clasp.

Confucius. Be aware of even doing that which you are likely, sooner or later, to repent for having done.

Baha'i. Be generous in prosperity and thankful in adversity.

Sophocles. Who is the slayer, who is the victim? Speak!

KINDNESS

Seneca. Wherever there is a human being, there is an opportunity for kindness.

Franklin Delano Roosevelt. Human kindness has never weakened the stamina or softened the fiber of a free people. A nation does not have to be cruel in order to be tough.

George Eliot (Mary Ann Evans). What do we live for, if it is not to make life less difficult for each other?

Christian Nestell Bovee. Kindness is a language the dumb can speak, and the deaf can hear and understand.

Johann von Goethe. Kindness is the golden chain by which society is bound together.

Octavius Brooks Frothingham.   Perfect obedience to the law of kindness would abolish government and the state.

Luc de Clapiers (Marquis de Vauvenargues).   We cannot be just unless we are kindhearted..

Seneca.   I had rather never receive a kindness, than never bestow one.

William Wordsworth.   The best portions of people's lives are their little, nameless, unremembered acts of kindness and love.

Samuel Smiles.   The cheapest of all things is kindness, its exercise requiring the least possible trouble and self-sacrifice.

Baha'i (Abdu'l-Bah'a).   The language of kindness is the lodestone of hearts and the good of the soul.

Plato.   Be kind, for everyone you meet is fighting a hard battle.

Marcus Aurelius.   Ask yourself daily, to how many ill-minded persons have you shown a kind disposition.

Washington Irving.   A kind heart is a fountain of gladness, making everything in its vicinity freshen into smiles.

Tulsi Das.   The root of all religions is kindness, the root of all sins is arrogance of power.

Blaise Pascal.   Kind words produce their own image in mortals' souls; and a beautiful image it is.

Samuel Johnson.   To cultivate kindness is a valuable part of the business of life.

Henry Charles Link.   If we were to make the conscious and frequent effort of treating others with consideration, the effects on us and on society as a whole would be amazing.

Charles Aleyn.   The true and noble way to kill foes, is not to kill them; you ,with kindness, may so change them that they shall cease to be foes.

George Eliot (Mary Ann Evans).   When death, the great reconciler, has come, it is never our tenderness that we repent of, but our severity.

French Proverb.   To speak kindly does not hurt the tongue.

Georges Bermamos.   Little things seem nothing, but they give peace, like those meadow flowers which individually seem odorless but all together perfume the air.

Science of Mind.   The understanding heart is filled with sympathy and helpfulness toward all.

Jack Kornfield.   When words are both true and kind, they can change our world.

Jean-Jacques Rousseau.   What wisdom can you find that is greater than kindness.

St. Augustine.   Those that are kind are free, though they are slaves; those that are evil are slaves, though they be kings.

Buddhism (Dalai Lama).   My religion is very simple. My religion is kindness.

Aldous Huxley.   Let us be kinder to one another.

Andre Gide.   True kindness presupposes the faculty of imagining as one's own, the suffering and joy of others.

Mother Teresa.   Kind words can be short and easy to speak, but their echoes are truly endless.

LIBERTY

Thomas Jefferson.   We hold these truths to be self-evident, that all people are created equal; that they are endowed by their creator with inalienable rights, and that among these are life, liberty and the pursuit of happiness.

Franklin Delano Roosevelt. The liberty of a democracy is not safe if the people tolerate the growth of private power to a point where it becomes stronger than their democratic state itself. That, in its essence, is fascism - ownership of government by an individual, by a group, or by any other controlling private power.

Dante Alighieri. The human race is in the best condition when it has the greatest degree of liberty.

Desmond (Archbishop) Tutu. We can be free only together....we can survive only together....we can be human only together.

Harold Joseph Laski. Without equality, I say, there cannot be liberty.

Henry Brooks Adams. Absolute liberty is absence of restraint; responsibility is restraint; therefore, the ideally free individuals are responsible to themselves.

William Wordsworth. We must be free or die, who speak the tongue that Shakespeare spoke, the faith and morals hold that Milton held.

Emily Bronte. And if I pray, the only prayer that moves my lips is "Leave the heart that now I bear and give me liberty!"

Pablo Neruda. Liberty sought you out in the mines, and begged for peace for your ploughs.

Immanuel Kant. What, therefore, can freedom possibly be but autonomy, that is, the property of the will to be a law to itself?

Albert Schweitzer. The highest possible material freedom for the greatest possible number is a requirement of civilization.

Edmund Burke. What is liberty without wisdom and without virtue?

Will Durant.   When liberty destroys order, the hunger for order will destroy liberty.

David Lloyd George. Liberty is not merely a privilege to be conferred; it is a habit to be acquired.

Earl Warren.   Liberty is the most contagious force in the world.

Mahatma Gandhi.   I am a lover of my own liberty and so would I do nothing to restrict yours.

Charles Caleb Colton.   Despotism can no more exist in a nation until the liberty of the press be destroyed.

John Fitzgerald Kennedy.   Unless liberty flourishes in all lands, it cannot flourish in one.

Nelson Mandela. Young people are capable, when aroused, of bringing down the towers of oppression and raising the banners of freedom.

Junius.   All despotism is bad; but the worst is that which works with the machinery of freedom.

John Dryden.   Oh, give me liberty! For even were paradise my prison, still I should long to leap the crystal walls.

LOVE

Buddhism.   Let a person cultivate towards the whole world a heart of love.

Christianity.   Beloved, let us love one another, for love is of God; and everyone that loves is born of God, and knows God. One that loves not, does not know God, for God is Love.

Samuel Taylor Coleridge.   The one prays best, who loves best all things both great and small; for the dear God who loves us, has made and loves all.

Confucianism.   To love all beings is the greatest benevolence.

164

Desmond Tutu. When you act lovingly you can begin to feel love.

Vincent Van Gogh. The best way to know God is to love many things.

Hinduism. One can best worship the Lord through love.

Marsha Sinetar. Only love heals, makes whole, takes us beyond ourselves. Love gets us there and lets us know who speaks.

Baha'i. (Baha'u'llah) O SON OF SPIRIT! My first counsel is this: Possess a pure, kindly and radiant heart.

Arthur Rubenstein. I have found that if you love life, life will love you back.

Islam. Love is this, that you should account yourself very little and God very great.

Johann Wolfgang Von Goethe. We are shaped and fashioned by what we love.

Taoism. Spirit arms with love those it would not see destroyed.

Dave Wilcox. In this scene set in shadows, like the night, love is here to stay. There is evil cast around us, but it's love that wrote the play.

Sikhism. God will regenerate those in whose hearts there is love.

Glenda Green. In its purest form, love is where creation creates itself. The key to love rests in knowing our own connection to that power.

Sidney Lanier. Music is love in search of a word.

Jainism. The days are of most profit to the one who acts in love.

Zoroastrianism. All beings are the beloved of the Lord and should love God in return.

Elizabeth Barrett Browning. Love me for love's sake, that evermore you may love on, through love's eternity.

Shinto. Love is the representative of the Lord.

Plato. All loves should be simply stepping-stones to the love of God.

Science of Mind (Ernest Holmes). Love is a cosmic force whose sweep is irresistible.

Mahatma Gandhi. Love is the prerogative of the brave.

De Lamennais. The heart of one who truly loves is a paradise on earth.

Kahlil Gibran. Love is the divine knowledge that enables one to see as much as the gods.

Mahatma Gandhi. What is obtained by love is retained for all time. What is obtained by hatred proves a burden in reality, for it increases hatred. The duty of a human being is to diminish hatred and to promote love.

Jack Kornfield. Love in the past is only a memory. Love in the future is a fantasy. Only here and now can we truly love.

June Jordan. One full black lily, luminescent in a homemade field of love.

Christianity (Paul the Apostle). So faith, hope and love abide, these three; but the greatest of these is love.

William James. I am done with great things and big plans and big success. I am for those tiny, invisible loving human forces that work from individual to individual.

Kahlil Gibran. Work is love made visible.

John Ruskin. Give a little love to a child, and you get a great deal back.

Christianity (Acts 18:10). I am with you, and no one shall try to hurt you.

MARRIAGE

Frederick William Robertson.   Marriage is not a union merely between two creatures - it is a union between two spirits, and the intention of that bond is to perfect the nature of both.

Kahlil Gibran.   You shall be together when the white wings of death scatter your days.

William Shakespeare.   God, the best maker of all marriages, combines your hearts in one.

Ovid.   If you would marry suitably, marry your equal.

Nicholas Rowe.   If you would have the nuptial union last, let virtue be the bond that ties it fast.

Henry David Thoreau.   There is more of good nature than of good sense at the bottom of most marriages.

George Eliot (Mary Ann Evans).   What greater thing is there for two human souls than to feel that they are joined for life.

Mahatma Gandhi.   Marriage is a natural thing in life, and to consider it derogatory in any sense is wholly wrong.

Honore de Balzac.   One should believe in marriage as in the immortality of the soul.

Kahlil Gibran.   Sing and dance together and be joyous, but let each one of you be alone, even as the strings of a lute are alone though they quiver with the same music.

Joseph Barth.   Marriage is our last, best chance to grow up.

Sir Thomas Beecham.   Marriage is an institution the appreciation of which increases as one grows older.

Joseph Addison.   A happy marriage has in it all the pleasures of friendship, all the enjoyments of sense and reason, and, indeed, all the sweets of life.

Mahatma Gandhi. The ideal is to look upon marriage as a sacrament and therefore to lead a life of self-restraint in the married state.

James Abram Garfield. The sanctity of marriage and the family relation make the corner-stone of our American society and civilization.

Henrik Ibsen. Marriage is something you have to give your whole mind to.

Samuel Johnson. Marriage is the strictest tie of perpetual friendship, and there can be no friendship without confidence, and no confidence without integrity.

William Congreve. Married in haste, we repent at leisure.

Charles Caleb Colton. That alliance may be said to have a double tie, where the minds are united as well as the body, and the union will have all its strength, when both the links are in perfection together.

Andre Maurois. A successful marriage is an edifice that must be rebuilt every day.

Channing Pollock. Marriage is the greatest educational institution on earth.

MYSTICISM

Albert Schweitzer. All deep religion, all profound philosophy, is in search for ethical mysticism and mystical ethics.

Kahlil Gibran. Those who define their conduct by ethics imprison their songbird in a cage.

John Donne No person is an island entire of itself. Any person's death diminishes me, because I am involved in humankind. Therefore never seek to know for whom the bell tolls; it tolls for you.

Jainism. Peace and universal love are the essence of the gospel preached by all the enlightened ones.

Eckhart Tolle Whatever "is" could not be otherwise.

Albert Einstein. The most beautiful thing we can experience is the mysterious. It is the source of all true art and science.

Boris Pasternak. What is laid down, ordered, factual, is never enough to embrace the whole truth.

Sam Keen. The holy is always experienced as a mystery that is at one and the same time awesome, majestic, and overpowering.

Robert Browning. The lark's on the wing; the snail's on the thorn; God's in His heaven - All's right with the world!

Edna St.Vincent Millay. I saw and heard, and knew at last the how and why of all things, past, and present, and forevermore.

William Blake. When the stars threw down their spears, and watered heaven with their tears, did God smile this work to see? Did the one who made the lamb make you?

Samuel Taylor Coleridge. Weave a circle round them thrice, and close your eyes with holy dread, for they on honey-dew have fed, and drunk the milk of paradise.

Sri Sri Ravi Shankar. The present is deep and vast. It is the secret of secrets.

Glenda Green. Our experience of the infinite is not a mystical ascension into some distant paradise, but a quiet and personal epiphany at moments when we realize that the miraculous and the mundane are one and the same.

Vedanta (Brahmavid). When the consciousness rises still higher, that which is the Reality shines, and we see it as the One Existence-Knowledge-Bliss, the Universal.

NEW PERSON AND CIVILIZATION

Nelson Mandela. The time for the healing of the wounds has come. The moment to bridge the chasms that divide us has come. The time to build is upon us.

Peter S. Adler. A new type of person whose orientation and view of the world profoundly transcends an indigenous culture is developing from the complex of social, political, economic, and educational interactions of our time.

Barbara Marx Hubbard. I believe that the transforming of the world is happening.

Winston Churchill. If the present tries to sit in judgment of the past, it will lose the future.

Tibetan Doctrine. Not to be excited by praise, not to be grieved by blame, but to know thoroughly one's own virtues or powers are the characteristics of an excellent person.

Baha'i (Baha'u'llah). Those virtues that befit dignity are forbearance, mercy, compassion, and loving-kindness towards all the peoples and kindreds of the earth.

John Dewey. Philosophy recovers itself when it becomes the method for dealing with the problems of humanity.

Shakti Gawain. Life is always challenging us to develop new aspects of ourselves and develop trust in new ways.

Taoism. The One that can be expressed is not the eternal One: The name that can be defined is not the unchanging name. Thought is called the antecedent of the universe; action is the mother of all things.

Baha'i. This is a new cycle of human power. It is the hour of unity among all peoples, and of the drawing together of all races and classes.

Hinduism. As one can ascend to the top of a house by means of a ladder or a bamboo or a staircase or a rope, so diverse are the ways and means of approaching the Divine, and every religion in the world shows at least one of these ways. .

Judaism. Let integrity and uprightness preserve me, for I wait on the Lord. God's radiance is a light to all the nations.

Christianity.   Blessed are the meek, for they shall inherit the earth. Blessed are the merciful, for they shall obtain mercy.

Islam.   The East and the West are God's: Truly God is immense, omnipresent, omniscient.

Zoroastrianism.   Seek to understand the forward movement of religion, and seek your share of duty therein.

Native American.   The Great Spirit above has appointed this place for us, on which to light our fires, and here we will remain. As to boundaries, the Great Spirit knows no boundaries, nor will his red children acknowledge any.

Louise Bogan.   Where is the shimmer of evil? This is the shell's iridescence and the wild bird's wing.

Sri Aurobindo. There will be established on earth a new Consciousness and Power which will shape a race of wise spiritual beings and take up into itself all of earth-nature that is ready for this new transformation.

John Dewey.   The measure of civilization is the degree in which the method of cooperative intelligence replaces the method of brute conflict.

Frank Lloyd Wright. Art and religion are the soul of our civilization. Go to them for there love exists.

Margaret Mead.   Never doubt that a small group of thoughtful, committed citizens can change the world. Indeed, it is the only thing that ever has.

NON-VIOLENCE

Mahatma Gandhi.   If people know God, they are incapable of harboring anger or fear within them no matter how overpowering the cause for it may be.

Nelson Mandela.   Government violence can do only one thing, and that is to breed counter violence.

Martin Luther King, Jr.   The non-violent approach gives (people) new self-respect; it calls up resources of strength and courage they did not know they had.

Desmond (Archbishop) Tutu.   Until women are deeply involved in opposing the violence in the world, we are not going to bring it to an end.

Marlo Thomas.   Not to kill another is what differentiates real people from mutated human creatures.

Louis Dembitz Brandeis.   As in the course of the heavenly bodies, harmony in national life is a result of the struggle between contending forces.

Thich Nhat Hanh.   Practice watering seeds of joy and peace and not just seeds of anger and violence, and the elements of war in all of us will be transformed.

Mahatma Gandhi.   Non-violence is the law of our species, as violence is the law of the brute. The dignity of humans requires obedience to such a higher law to strengthen the Spirit.

Baha'i (Baha'u'llah).   The sword of a virtuous character and upright conduct is sharper than blades of steel.

Martin Luther King, Jr.   Returning violence for violence multiplies violence, adding deeper darkness to a night already devoid of stars. Darkness cannot drive out hate; only love can do that.

Mahatma Gandhi.   Non-violence is based on the assumption that human nature...unfailingly responds to the advances of love.

A Course in Miracles.   I can escape the world I see by giving up attack thoughts.

Emma Goldman.   It is organized violence on top which creates individual violence at the bottom.

A Course in Miracles.   Holding grievances is an attack on God's plan for salvation.

Elizabeth I. I will never be by violence constrained to do anything.

Jack Kornfield. Our own worst enemy cannot harm us as much as our unwise thoughts. Nothing can help us as much as our own compassionate thoughts.

Simone de Beauvoir. Woman is in any case deprived of the lessons of violence by her nature.

Jack Kornfield. Even loss and betrayal can bring us awakening.

John Bright. Force is not a remedy.

John Milton. Who overcomes by force, overcomes only half a foe.

Mahatma Gandhi. Human nature is so constituted that if we take absolutely no notice of anger or abuse, the person indulging in it will weary of it and stop.

Seneca. Power exercised with violence has seldom been of long duration, but temper and moderation generally produce permanence in all things.

Dennis J. Kucinich. Let us work to make non-violence an organizing principle in our own society.

ONENESS

Po Chu-I. The bond that joined us lay deeper than outward things: the rivers of our souls spring from the same well.

Morihei Ueshiba. The universe emerged and developed from one source, and we evolved through the optimal process of unification and harmonization.

Judaism. Hear, O Israel; the Eternal is our God, the Eternal is one. The Eternal One of All Being is the God within us.

Baha'i. You are all the leaves of one tree and the drops of one ocean.

Buddhism. What is meant by the soul of suchness is the oneness of the totality of things, the great all-including whole. For the essential nature of the soul is eternal.

Christianity. There is one body, and one Spirit, even as you are called in one hope of your calling; one Lord, one faith, one God who is above all, and through all, and in all.

Vedanta (Sri Ramakrishna). The sunlight is one and the same wherever it falls, but only bright surfaces like water, mirrors and polished metals can reflect it fully. So is the Divine Light. It falls equally and impartially on all hearts, but only the pure and clean hearts of the good and holy can fully reflect it.

Islam. The East and the West are God's; therefore, whichever way you turn, there is the face of God. Truly God is immense, knowing.

Taoism. We who know what God is, and who know what humanity is, have attained.. Knowing what humanity is, we rest in the knowledge of the unknown. Working out one's allotted span and not perishing in mid career, this is the fullness of knowledge.

Athenagoras. Sometimes I feel that I belong to all the religions.

Albert Einstein. All religions, arts and sciences are branches of the same tree.

Mahatma Gandhi. It is on soul-force or spiritual force that the true religious reformer has hitherto relied.

George Bernard Shaw. There is only one religion, though there are a hundred versions of it.

Voltaire. The tolerance of all religions is the law of nature stamped on the hearts of mortals.

Shirley Leighton. All is one in its center - no judgement denies that birthright. No resistance is necessary - only acknowledgment of the one core of being inherent in all.

Vedanta (Swami Vivekananda). Every religion is evolving a God out of the material realm, and the same God is the inspirer of all of them. (Parliament of the World's Religions - 1893).

OUR PLANET, EARTH

Mary Webb. Love me, and I will give into your hands, the rare, enameled jewels of my lands, and when, like a passing light along the sky, your wild-bird soul shall clap her wings and fly, she shall but nest more closely in my breast.

Buddhism. So within yourself foster a limitless concern for every living creature.

Teilhard de Chardin. I live at the heart of a single, unique Element, the Center of the Universe, and am present in each part of it; personal love and cosmic power.

Hinduism. Waters, you are the ones who bring us the Life Force. Help us to find nourishment so that we may look upon great joy.

Gerard de Nerval. Each flower is a soul opening out to nature.

Islam. The totality of the environment is God's creation and our responsibility to protect.

Native American. Make us wise so that we may understand what you have taught us. Help us to learn the lessons you have hidden in every leaf and rock.

Religious Society of Friends. Do I love simply, mindful how my life affects the earth and its resources?

Baha'i. It is not only their fellow human beings that the beloved of God must treat with mercy and compassion; rather must they show forth the utmost loving-kindness to every living creature.

Lord Byron. Are not the mountains, waves and skies a part of me and of my soul, as I of them?

Sikhism. Nature is not only the source of life, beauty and power, but it is also an inspiration of strength in formulation of character.

Alan Watts. Every individual is an expression of the whole realm of nature, a unique action of the total universe.

Martin Luther King, Jr. Truth pressed to the earth rises again.

Robert Kennedy. The resources of the earth and the ingenuity of mortals can provide abundance for all.

Paul Bigelow Sears. The good earth is our mother and if we destroy her, we destroy ourselves.

Sioux Indian. Behold this and always love it! It is very sacred, and you must treat it as such.

Maori. The land is a mother that never dies.

PEACE

Dennis J. Kucinich. Let us recommit ourselves to the slow and painstaking work of statecraft, which sees peace, not war as being inevitable.

Morihei Ueshiba. One does not need buildings, money, power, or status to practice the Art of Peace. Heaven is right where you are standing, and that is the place to train.

John Marsden. May the knife remain in the holder, may the bullet stay in the gun, may those who live in the shadows be seen by those in the sun.

Native American (Tsalagi). Affirm that peace may prevail through complimentary resolution.

John Milton. Peace has her victories no less renowned than war

Christianity. Blessed are the peacemakers, for they shall be called the children of God.

Dalai Lama. Peace, in the sense of the absence of war, is of little value to someone who is dying of hunger or cold...Peace can only last where human rights are respected, where the people are fed, and where individuals and nations are free.

Cecil Spring-Rice. And soul by soul, and silently her shining bounds increase; her ways are ways of gentleness and all her paths are peace.

Confucianism. Seek to be in harmony with all your neighbors. Live in PEACE with all beings.

Baha'i (Baha'u'llah). Yet so it shall be; these fruitless strifes, these ruinous wars shall pass away, and the 'Most Great Peace' shall come.

Buddhism. There is no happiness greater than peace.

Anonymous. Peacemakers are those who can keep their own integrity while seeing understanding and having compassion for the conflicting sides in a dispute.

Barbara Lee. Peace is patriotic.

Hinduism. Without meditation, where is peace? Without peace, where is happiness?

Pablo Neruda. Liberty sought you out in the mines, and begged for peace for your ploughs.

Islam. God will guide us to peace. If we will heed God's words, we will be led from the darkness of war to the light of peace.

Taoism. The wise esteem peace and quiet above all else.

Sikhism. Only in the Name of the Lord do we find our peace.

Judaism. When people's ways please the Lord, even their enemies come to be at peace with them.

Jainism. All beings should live in peace with their fellows. This is the Lord's desire.

Zoroastrianism. I will sacrifice to peace, whose breath is friendly.

Baha'i. War is death while peace is life.

Shinto. Let the earth be free from trouble and people live at peace under the protection of the Divine.

Albert Einstein. Peace cannot be kept by force. It can only be achieved by understanding.

Mahatma Gandhi. Each one has to find peace from within. Peace to be real must be unaffected by outside circumstances.

Ralph Waldo Emerson. Nothing can bring you peace but yourself.

John F. Kennedy. Peace is a daily, a weekly, a monthly process.

Dorothy Thompson. They have not wanted Peace at all; they have wanted to be spared war - as though the absence of war was the same as peace.

Anonymous. 11 year old. Peace is a special thought or a special love or light or spark that we all share within ourselves.

Thich Nhat Hanh. If in our daily life, we can smile...not only we, but everyone will profit from it. This is the most basic kind of peace work.

St. Theresa of Lizieux, Let us not be justices of peace, but angels of peace.

Dag Hammarskjold. There is no peace which is not peace for all; no rest till all has been fulfilled.

Anne O'Hare McCormick. The responsibility for making peace rests, above all, on the country that is most nearly normal as the war ends.

Petrarch. Five great enemies to peace inhabit with us: viz., avarice, ambition, envy, anger, and pride. If those enemies were to be banished, we should infallibly enjoy perpetual peace.

U Thant. I am convinced that if peace were soon restored there would be a rebirth of faith in our ability to promote the well-being of all.

James Thomson. Peace is the happy, natural state of humans; war, their corruption, their disgrace.

Epicurus. Only the just person enjoys peace of mind.

Makim Maksimovich Litvinov. Peace is indivisible.

Jawaharlal Nehru. Peace cannot suddenly descend from the heavens. It can only come when the root-causes of trouble are removed.

Raymond Gram Swing. The alternative to peace is not war. It is annihilation.

Marianne Moore. The world's an orphans' home. Shall we never have peace without sorrow?

Adlai Ewing Stevenson. Peace is the one condition of survival in this nuclear age.

Stephen Longfellow Fiske. We are important and valued members of the family of humanity. We are contributors to peace.

Thich Nhat Hanh. To prevent the next war, we have to practice peace today. If we establish peace in our hearts and in our ways of looking at things, war will not come. If we wait until another war is imminent to begin to practice it, it will be too late.

Navajo Indian. Before me peaceful, behind me peaceful, under me peaceful, over me peaceful, all around me peaceful.

PERCEPTION AND ATTITUDE

Ralph Waldo Emerson. Nothing can bring you peace but yourself.

Ethel Barrymore. You grow up the day you have your first real laugh at yourself.

Baruch Spinoza. The more clearly you understand yourself and your emotions, the more you become a lover of what is.

Jack Kornfield. The heart is like a garden. It can grow compassion or fear, resentment or love. What seeds will you plant there?

Patricia Sun. There is only one thing stopping us from having heaven on earth: that we can't believe it can be.

W. H. Murray. The moment one definitely commits oneself, then Providence moves too.

M. P. Follett. Concepts can never be merely presented. They must be knitted into the structure of my being. This can only be done through my own activity.

Gurumayi. Approach the present with your heart's consent. Make it a blessed event.

Eckhart Tolle. When you surrender to what is and so become fully present, the past ceases to have any power.

Marcel Proust. The real act of discovery consists not in finding new lands but in seeing with new eyes.

Richard Bach. You are never given a wish without also being given the power to make it true. However, you may have to work for it.

John Fitzgerald Kennedy. Peace is a daily, a weekly, a monthly process.

Shirley Leighton. Ideals furnish the building blocks with which we form our construct of the world in which we choose to live.

Jack Kornfield. If you can't find the truth right where you are, where else do you think you will find it?

Anonymous. Control your attitude and emotions, or they will control you.

Walt Whitman.   Who makes much of a miracle? As for me, I know of nothing else but miracles.

Anne Wilson Schaef.   Trust the unfolding of your life and see what you have to learn from that.

Eckhart Tolle.   Suffering is necessary until you realize it is unnecessary.

Henry James.   It's time to start living the life you've imagined.

Albert Einstein.   In the middle of a difficulty lies opportunity.

Lynn Andrews.   What we need is the courage to see that everything around us in our lives is a mirror.

Judaism (Hillel).   If I am not for myself, who will be for me? If I am only for myself, what am I? And if not now, when?

Edith Wharton.   There are two ways of spreading light - to be the candle or the mirror that reflects it.

SACREDNESS

Native American (Black Elk).   Some little root of the sacred tree still lives. Nourish it, that it may leaf and bloom and fill with singing birds.

Alfred Lord Tennyson.   Self-reverence, self-knowledge, self control: These three alone lead life to sovereign power.

Emma Curtis Hopkins.   To love is to see God in all.

Joel S. Goldsmith.   God is the only law, a law which maintains and sustains the harmony and perfection of its own creation at all times.

Christien de Quincey.   Consciousness is the self creation and self organization of matter and energy. In other words, all matter is ensouled.

Hinduism (Bhagavad Gita).   Only that yogi whose joy is inward and whose vision inward shall come to Brahman and know Nirvana.

Shirley Leighton.   When we love enough, we can never be hurt because love and light overpower all negatives.   So love everything - especially what you think you don't want.

Judaism (Jeremiah 33:3).   Call to me, and I will answer you, and show you great and mighty things, which you do not know.

Riane Eisler.   What we think of as 'sacred' actually is present in everything we do.

Christianity (Matthew 6:22).   The light of the body is the eye.   If therefore your eye is single, your whole body shall be full of light.

Zorastrianism.   Life would be intolerable if it were not for the sympathy, kindness, and affection of person to person.

Buddhism (Al Duffy).   Those who transform greed, anger, and lust into generosity, equanimity and righteousness reside in a heavenly realm.

Glenda Green.   What is common to all sacred moments is surrender of the human condition to possibilities beyond estimation and comprehension.

SCIENCE AND SPIRITUALITY

Aristotle.   Philosophy is the science which considers truth.

Brooke Medicine Eagle.   Spirit lives in you; it lives within your body. You can touch the Great Spirit by touching into your own aliveness.

Albert Einstein.   The most beautiful emotion we can experience is the mystical.   It is the power of true art and science.   Those to whom this emotion is a stranger, who can no longer wonder and stand rapt in awe, are as good as dead.

Jack Kornfield.   Life is as fleeting as a rainbow, a flash of lightning, a star at dawn. Knowing this, how can we quarrel?

Edmund W. Sinnott.   Life is the center where the material and spiritual forces of the universe seem to meet and to be reconciled.   Spirit is born in life.

Alfred North Whitehead. Philosophy is the attempt to make manifest the fundamental evidence as to the nature of things.

Deepak Chopra. Look at your physical body as a printout of signals being sent back and forth between your brain and every cell.

Duane Elgin. Voluntary simplicity is not so much about living with less as it is about living with purpose and balance.

Vedanta (Swami Vivekananda). We want that bright sun of intellectuality, joined with the heart of Buddha, the wonderful, infinite heart of love and mercy.

Eckhart Tolle. A great silent space holds all of nature in its embrace. It also holds you.

Emily Dickinson. The soul should always stand ajar, ready to welcome the ecstatic experience.

Georgia O'Keefe. If you take a flower in your hand and really look at it, it's your world for a moment.

Michael Guillen. Precisely because one aspect of our behavior is chaotic, is unpredictable, the seemingly impossible will always be probable.

Pierre Simon de Laplace. In the small number of things we are able to know with any certainty, the principal means of ascertaining truth is based on probabilities.

SERVICE

Jelaluddin Rumi. Be a lamp, or a lifeboat, or a ladder. Help someone's soul to heal. Walk out of your house like a shepherd.

Thich Nhat Hanh. If in our daily life, we can smile... not only we, but everyone will profit from it. This is the most basic kind of peace work.

Mahatma Gandhi. To forget how to dig the earth and tend the soil is to forget ourselves.

Mother Teresa.   Find those who think they are alone and let them know that they are not.

Martin Luther King, Jr. Everybody can be great....because anybody can serve...You only need a heart full of grace and a soul generated by love.

Hinduism (Lord Krishna). Do your duty and do not expect rewards: they are already destined for the action you do.

Hans Selve. A long, healthy and happy life is the result of making contributions, of having meaningful projects that are personally exciting and contribute to the lives of others.

Baha'i (Abdu'l-Bah'a).   Is there any deed in the world that would be nobler than service to the common good?

Josephine Kermode.   Yours is the work to save these sheep, your glory let it be; for every soul in Cornadale, you, John, will answer Me!

Jack Kornfield.   The more fully we give our energy, the more it returns to us.

Mother Teresa.   The fruit of love is service, which is compassion in action.

Benjamin Disraeli.   The greatest good you can do for others is not just to share your riches, but to reveal to them their own.

Ralph Waldo Emerson. It is one of the most beautiful compensations of life that people cannot sincerely try to help another without helping themselves.

Judaism (Ecclesiastes).   Keep on sowing your seeds, for you never know which will grow - perhaps they all will.

Jack Kornfield.   We do not possess our home, our children, or even our own body. They are only given to us for a short while to treat with care and respect.

Native American (Bear Paw). We are here to serve nature in all its creativity and procreation. All we need to do is to let our veins flow naturally from the moment of life. Don't let anything impede your natural instinct to flow in all directions.

Henry David Thoreau. Be not simply good; be good for something.

Robert Browning. All service is the same with God.

Mahatma Gandhi. Service without humility is selfishness and egotism.

Sir Wilfred Thomason Grenfell. The service we render to others is really the rent we pay for our room on this earth.

STILLNESS, PRAYER AND MEDITATION

Thich Nhat Hanh. Breathing in, I know that I am breathing in. Breathing out, I know that I am breathing out.

Science of Mind. A great stillness steals over me and a great calm quiets my whole being, as I realize Your Presence.

Agnes de Mille. No trumpets sound when the important decisions of our life are made. Destiny is made known silently.

Vedanta (Swami Brahmavidyanhanda). The practice of meditation is the great scientific method of attaining knowledge.

Kahil Gibran. Love is the divine knowledge that enables one to see as much as the gods.

Christianity (Matthew). Where two or more are gathered in my name, there am I in the midst of them.

Sri Sri Ravi Shankar. Meditation works by bringing an effect from the level of Being to the mind. With the breath we bring this effect to the physical level as well.

Judaism (Proverbs 3:6). Acknowledge God in all your ways, and Spirit shall direct your paths.

Native American (Bear Paw). A blank stare focusing on any object conducts a stillness and relaxation of mind and body, which releases dreaming states of prayer. Sit in nature and look beyond.

Baha'i. The faculty of meditation is the depository of crafts, arts and sciences. Exert yourselves, so that the gems of knowledge and wisdom may proceed from this ideal mine, and conduce to the tranquility and union of the different nations of the world.

Kahlil Gibran. They alone are great who turn the voice of the wind into a song made sweeter by their own loving.

Vedanta (Swami Brahmavidyananda). People may never have entered a church or a mosque, nor performed any ceremony, but if they feel God within themselves and are thereby lifted above the vanities of the world, those people are holy.

Mahatma Gandhi. I believe that prayer is the very soul and essence of religion, and therefore prayer must be the very core of the life of a person. Begin therefore, your day with prayer and close your day with prayer so that you may have a peaceful night.

Eckhart Tolle. When you lose touch with inner stillness, you lose touch with yourself. When you lose touch with yourself, you lose yourself in the world.

Baha'i. The nightingale of the rose garden of uprightness must display its wonderful melodies and trills.

Konko Daijin. Living spirit of our Principal Parent of the universe, pray with a single heart. The divine favor depends upon one's own heart. On this very day, pray.

Vedanta (Swami Brahavidyananda). It is meditation that brings us nearer to truth than anything else.

Eckhart Tolle. Nature can bring you to stillness. That is its gift to you. When you perceive and join with nature in the field of stillness, that field becomes perforated with your awareness. That is your gift to nature.

STRENGTH

Taoism (Lao Tze). By the accident of fortune, people may rule the world for a time, but by virtue of love and kindness, they may rule the world forever.

Eleanor Roosevelt. Accept whatever comes; the only important thing is that you meet it with courage and with the best you have to give.

Mahatma Gandhi. Strength does not come from physical capacity. It comes from an indomitable will.

Friedrich Wilhelm Nietzsche. One who has a WHY to live can bear almost any HOW.

Ralph Waldo Emerson. It is easy in the world to live after the world's opinion; it is easy in solitude to live after our own; but the great person is the one who, in the midst of the crowd, keeps with perfect sweetness the independence of solitude.

Peter Marshall. When we long for life without difficulties, remind us that oaks grow strong in contrary winds and diamonds are made under pressure.

Jack Kornfield. To open our own heart like a Buddha, we must embrace the ten thousand joys and the ten thousand sorrows.

Anonymous. Heroes are the people who do what has to be done when it needs to be done regardless of the consequences.

N. Sri Ram. Only as we go out in love which seeks to help and serve do we transcend ourselves and develop that consciousness which embodies the awareness of our essential unity with others.

Mahatma Gandhi. God is an unseen power residing within us.

Christianity (Luke 22:27). I am among you as one who serves.

Richard Bach. The mark of your ignorance is the depth of your belief in injustice and tragedy. What the caterpillar calls the end of the world, the master calls a butterfly.

Judaism (Isaiah 30:15). In quietness and in confidence shall be your strength.

American Indian (Tsalagi).. To actualize is to manifest the ideal through sacred practice, great diligence, perseverance, and perspiration.

Jack Kornfield. Whatever we cultivate in times of ease, we gather as strength for times of change.

William Hazlitt. Those who can command themselves command others.

Air Force Motto. The difficult we do immediately. The impossible takes a little longer.

Kahlil Gibran. Tenderness and kindness are not signs of weakness and despair but manifestations of strength and resolution.

TRANSCENDENT HUMANITY

Maria Mitchell. Long before the dawn of another centennial the struggle for the equality of the sexes will have given place to higher and nobler issues for the advancement of humanity.

Islam (Qur'an) God does not look at your bodies and figures but looks at your hearts and deeds..

Jelaluddin Rumi (Quatrain 158). Out beyond ideas of wrongdoing and right doing, there is a field. I'll meet you there.

Mahatma Gandhi. For I can see that in the midst of death, life persists; in the midst of untruth, truth persists; in the midst of darkness, light persists. Hence, I gather that God is life, truth, light. God is love. God is the supreme good.

Jean Shinoda Bolen. My deep sense of spiritual meaning has to do with remembering as much as it has to do with discovering.

Sogyal Rinpoche. Pain, grief, loss, and ceaseless frustration of every kind are there for a real and dramatic purpose: to wake us up...and...release our imprisoned splendor.

Martin Luther King, Jr.   I believe that unarmed truth and unconditional love will have the final word in reality. This is why right, temporarily defeated, is stronger than evil triumphant.

Religious Science (Ernest Holmes).   Like the creative soil in which seeds are planted, the Soul of the Universe is the Creative Medium into which the Word of Spirit falls and from which Creation rises.

Terra Ryan.   Adversity does not build character; it reveals it.

Bawa Muhaiyadden.   Trust in Allah, and the path will open.

Thich Nhat Hanh.   Practice watering seeds of joy and peace and not just seeds of anger and violence, and the elements of war in all of us will be transformed.

Managing From the Heart.   Hear and understand me. Even if you disagree, please don't make me wrong. Acknowledge the greatness within me. Remember to look for my loving intentions. Tell me the truth with compassion.

N. Sri Ram.   Only as we go out in love which seeks to help and serve do we transcend ourselves and develop that consciousness which embodies the awareness of our essential unity with others.

Dr. William Hornaday. The highest pinnacle rises from the broadest base.

Sri Sri Ravi Shankar.   What is the meaning of life? Ah...this you better find out for yourself.

Albert Schweitzer.   My happiness will be incomplete as long as one creature is miserable.

Taoism (Lao-Tze).   He who knows others is wise; he who knows himself is enlightened.

Jane Roberts.   Dreams are the mind's free play. The spontaneous activity however, is at the same time training in the art of forming practical events.

Bawa Muhaiyaddeen. Place the truth in front of you and follow it. It will reveal to you the path of the liberation of the soul.

John F. Kennedy. Our most basic common link is that we all inhabit this planet. We all breathe the same air. We all cherish our children's future. And we are all mortal.

Baha'i. Be fair in judgement and guarded in your speech

Mahatma Gandhi. God is an unseen power residing within us.

Alfred Lord Tennyson. Self reverence, self-knowledge, self- control: These three alone lead life to sovereign power.

Ralph Waldo Emerson. It is easy in the world to live after the world's opinion; it is easy in solitude to live after our own; but the greatest are those who, in the midst of the crowd keep with perfect sweetness the independence of solitude.

Joseph Addison. What sunshine is to flowers, smiles are to humanity. They are but trifles, to be sure; but scattered among life's pathway, the good they do is inconceivable.

Christianity (Matthew 6:32). Which of you by taking thought can add one cubit to your stature?

TRUTH

Sun Tzu. Deception is for the purpose of seeking victory over an enemy; to command a group requires truthfulness.

Anne Sophie Swetchine. When two truths seem directly opposed to each other, we must not question either, but remember there is a third - God - who reserves the right to harmonize them.

Martin Luther King, Jr. I believe that unarmed truth and unconditional love will have the final word in reality.

Dalai Lama. My hope rests .... in the love of truth and justice which is still in the heart of the human race.

Richard Bach. Here is a test to find whether your mission on earth is finished. If you are alive, it isn't.

Science of Mind. There is one spirit back of all expression.

Ayfo Onee. Where is my light? My light is in me. Where is my hope? My hope is in me. Where is my strength? My strength is in me - and in you.

Christian D. Larso. A genius is asleep in the subconscious of every mind; a spiritual giant is within us awaiting recognition; and in the soul is the Christ knocking at the door.

Jack Kornfield. At the bottom of things, most people want to be understood and appreciated.

Nietzsche. He who has a WHY to live can bear 'most any HOW.

Emily Dickinson. Truth is as old as God, a twin identity - and will endure as long as God.

Mahatma Gandhi. An eye for an eye makes the whole world blind.

Vedanta (Sri Ramakrishna). Those who cling tenaciously to truth ultimately realize God. Without this regard for truth, one gradually loses everything.

Judaism (Ecclesiastes 11:1). Cast your bread upon the waters and you shall find it after many days.

Science of Mind (Ernest Holmes). To desert the truth in the hour of need is to prove that we do not know the truth.

Linus Pauling. Satisfaction in one's curiosity is one of the greatest sources of happiness in life.

Zorastrianism. Truth knows no racial or geographical boundaries. Truth is ever the same for all. Truth builds and creates; falsehood breaks and destroys.

Henri Amiel. Truth is not only violated by falsehood; it may be equally outraged by silence.

Ralph Waldo Emerson. The greatest homage we can pay to truth is to use it.

Andre Gide. To love the truth is to refuse to let oneself be saddened by it.

Harold Le Claire Ickes. The path of truth is the path of progress.

UNITY-AND-DIVERSITY

Amy Lowell. What if all religions be true, and Gabriel's trumpet blows for you and blows for them - what will you do?

Islam. And of God's signs is the creation of the heavens and the earth, and the difference of your languages and colors. Lo! Herein is deed over potents for people of knowledge.

Fatema Mernissi. We can bring a new world into being through all the scientific advances that allow us to communicate, to engage in unlimited dialogue, to create that global mirror in which all cultures can shine in their uniqueness.

Baha'i. The diversity in the human family should be the cause of love and harmony, as it is in music where many different notes blend together in the making of a perfect chord.

Nelson Mandela. What challenges us is to ensure that none should enjoy lesser rights; and none tormented because they are born different, hold contrary political views, or pray in a different manner.

Mahatma Gandhi. I recognize truth by the name of Rama. "This is my way even though I honor yours".

Ruth Pitter. Thrust back contention, merge in one, warring dualities; make free, night of the moon, day of the sun; end the old war of land and sea.

Group dialogue at Eco Village, 1998.   The result of a search for wholeness in which separateness, uniqueness and particularity are enhanced by the realization that "we are one".

Jack Kornfield.   We are not (just) independent but interdependent.

Christianity (Philippians 4:8).   Whatsoever things are true, whatsoever things are honorable, whatsoever things are just, whatsoever things are pure, whatsoever things are lovely, whatsoever things are of good report; if there be any virtue, and if there by any praise, think on these things.

Judaism (Ecclesiastes 9:17).   The words of wise people are heard in quiet more than the cry of those who rule among fools.

Vedanta (Swami Vivekananda).   Humanity must progress as a whole. Unity in variety, not uniformity, is the pattern for world-culture. There is no inherent conflict between science and religion, between reason and faith, or between poetry and philosophy.

Elizabeth Kubler-Ross.   Learn to get in touch with the silence within yourself and know that everything in this life has a purpose.

Albert Einstein.   In the middle of difficulty lies opportunity.

Sam Keen. On their surface the world's religions are irreconcilably diverse; but at their core, in their experience of the holy, they are in near universal agreement.

Jean-Jacques Rousseau.   The world of reality has its limits; the world of imagination is boundless.

Irish Proverb.   It is in the shelter of each other that the people live.

Vedanta (Swami Vivekananda).   Our watchword will be acceptance and not exclusion. We believe in acceptance.

Teilhard de Chardin.   Between the within and the without of things, the interdependence of energy is incontestable.

Zora Neale Hurston. I have no separate feeling about being an American citizen and colored. I am merely a fragment of the Great Soul that surges within the boundaries.

WISDOM

Cicero. Philosophy, if rightly defined, is nothing but the love of wisdom.

Brahma Kumaris Our original religion is peace.

Native American (Bear Paw). Wisdom is listening attentively while the words of another spill forth. It is letting others use your silence to vent their thoughts.

Baha'i (Baha'u'llah). The sword of wisdom is hotter than summer heat, and sharper than blades of steel.

William Shakespeare. You should not be old until you have been wise.

Anne Frank. We all live with the objective of being happy; our lives are all different, and yet the same.

Bawa Muhaiyaddeen. Do not waste your time in useless talk. Silently learn to speak the wisdom which exists within speech. That will give you tremendous peace.

Martin Luther. God writes the gospel not in the Bible alone, but on trees, and flowers, and clouds, and stars.

Judaism (Ecclesiastes 3:12). To everything there is a season, and time for every purpose under the sun; a time to be born, a time to die; a time to plant and a time to reap.

Science of Mind (Ernest Holmes). To be still and know that the Eternal Presence is in us is the beginning of wisdom and of freedom.

Jalaluddin Rumi. Why do you stay in prison, when the door is so wide open?

Science of Mind.   There is no power but ourselves in the universe that can free us.

Bawa Muhaiyadden.   With wisdom analyze the sadness and the wrong thoughts that enter your mind and then throw them away. Try to do what is good.

Jack Kornfield. We do not need more knowledge but more wisdom. Wisdom comes from our own attention.

Judaism (Psalms 11:10).   The fear of the Lord is the beginning of wisdom: those who practice it have good understanding.

Shirley Leighton.   One begins to realize that every spark is made of the ONE LIGHT.

Jack Kornfield.   If you are poor, live wisely. If you have riches, live wisely. It is not your station in life but your heart that brings blessings.

Richard Bach.   There is no such thing as a problem without a gift for you in its hands. You seek problems because you need their gifts.

Ralph Waldo Emerson.   To make knowledge valuable, I must have the cheerfulness of wisdom.

Christianity (Matthew 7:6).   Do not throw your pearls before swine lest they trample them underfoot and turn to attack you.

Buddhism (Bodhidha-ma).   When the world and the mind are both transparent, this is the true vision. And such understanding is true understanding.

Confucianism. Of neighborhoods, benevolence is the most bountiful. How can people be considered wise who when they have the choice, do not settle on benevolence?

Marcus Aurelius.   Live not one's life as though one had a thousand years, but live each day as the last.

Marie Curie.   Nothing in life is to be feared. It is only to be understood.

Edward Fitzgerald. The Moving Finger writes; and, having written, moves on; nor all your piety nor wit shall lure it back to cancel half a line, nor all your tears wash out a word of it.

WOMAN

Julian of Norwich. To the property of motherhood belong nature, love, wisdom and knowledge, and this is God.

Alice Meynell. She walks - the lady of my delight - a shepherdess of sheep. Her flocks are thoughts, she keeps them white; she guards them from the steep.

United Nations. The full and complete development of the world and the cause of peace requires the maximum participation of women as well as men in all fields.

Carol Gilligan. Sensitivity to the needs of others and the assumption of responsibility for taking care lead women to attend to voices other than their own and to include in their judgment other points of view.

William Shakespeare. It is beauty that often makes women proud; it is virtue that makes them most admired; it is government that makes them seem divine.

Desmond (Archbishop) Tutu. Unleashing the power of women has the potential to transform our world in extraordinary and many as yet unimagined ways.

Elinor Wylie. I was, being human, born alone; I am, being woman, hard beset; I live by squeezing from a stone the little nourishment I get.

Mahatma Gandhi. If society is not to be destroyed by insane wars of nations against nations, and still more insane wars on its moral foundations, the woman will have to play her part.

Simone de Beauvoir. When we abolish the slavery of half of humanity, together with the whole system of hypocrisy that it implies, then the "division" of humanity will reveal its genuine significance and the human couple will find its true form.

Anne Morrow Lindberg.   By and large, mothers and house-wives are the only workers who do not have regular time off. They are the great vacationless class.

Betty Friedan.   A girl should not expect special privileges because of her sex, but neither should she "adjust" to prejudice and discrimination. She must learn to compete...not as a woman, but as a human being.

Susan B. Anthony.   Woman must not depend upon the protection of man, but must be taught to protect herself.

Elizabeth Cady Stanton.   We hold these truths to be self-evident, that all men and women are created equal.

Sojourner Truth.   I have plowed, and planted, and gathered into barns, and no man could head me - and arn't I a woman? I could work as much and eat as much as a man (when I could get it), and bear the lash as well - and arn't I a woman?

William Shakespeare.   Age cannot wither her, nor custom stale her infinite variety.

Muriel Rukeyser.   When your women are ready and rich in their wish for the world, destroy the leaden heart, we've a new race to start.

Margaret Fuller.   It is a vulgar error that love, *a* love, to Woman is her whole existence; she also is born for Truth and Love in their universal energy.

Eleanor Roosevelt.   Too often the great decisions are originated and given form in bodies made up wholly of men, or so completely dominated by them that whatever of special value women have to offer is shunted aside without expression.

Louise Renne.   What is enough? Enough is when somebody says, "Get me the best people you can find" and nobody notices when half of them turn out to be women.

Desmond Tutu.   We have hardly spoken about the Motherhood of God, and consequently we have been the poorer for this.

RESPONSIVE

READINGS

CONTENTMENT, Tao-Te-Ching

In dwelling, think one's place suitable: in feeling, make the heart deep;

In friendship, keep on good terms with all; in words, have confidence;

In ruling, abide by good order; in business, take things easy;

In motion, make use of the opportunity. Since there is no contention, there is no blame.

Seeking and holding material gains to the very end - it is better to leave them alone;

Handling and sharpening a weapon - it cannot be long left unused;

When gold and jade fill the hall, no one can protect them;

Wealth and honor with false pride bring with them destruction;

Having given your best, then withdraw into obscurity -

This is the way to the Eternal.

Fame or your conscience, which is nearer to you?

Your conscience or wealth, which is dearer to you?

Gain or loss, which brings more evil to you?

Over-love of anything will lead to wasteful spending:

Amassed riches will be followed by heavy plundering.

Therefore, those who know contentment (with few possessions) can never be humiliated;

Those who know where to stop can never be perishable:

They will long endure.

When the Eternal is realized in the world, the swift horses are used for doing work in the field.

When the Eternal is not realized in the world, war horses are bred on the commons outside the cities.

There is no greater crime than knowing no content:

There is no greater calamity than indulging in greed.

Therefore, the sufficiency of contentment is an enduring and unchanging sufficiency.

EIGHTFOLD PATH, Buddhism

What, friends, is the truth concerning the way that leads to the cessation of ill?

This is the eightfold path; right view, right aspiration, right speech, right doing, right livelihood, right effort, right mindfulness, right rapture.

And what, friends, is right view?

Knowledge about ill, knowledge about the coming to be of ill, knowledge about the cessation of ill, knowledge about the way that leads to the cessation of ill.

And what, friends, is right aspiration?

The aspiration towards renunciation of the fruits of action, the aspiration towards benevolence, the aspiration towards kindness.

And what, friends, is right speech?

Abstaining from lying, slander, abuse and idle talk.

And what, friends, is right doing?

Abstaining from taking any life unnecessarily, from taking what is not given, from carnal indulgence.

And what, friends, is right livelihood?

The good disciples, having put away wrong livelihood, support themselves by right livelihood, which is dictated by their nature.

And what, friends, is right effort?

Here individuals make effort in bringing forth will-power, so that evil states that have not arisen within them may not arise; to that end they call forth energy and determination.

And what, friends, is right mindfulness?

Herein individuals, as to the body, continue so to look upon the body, that they remain ardent, self- assessed and mindful, having overcome both the hankering and the dejection common in the world.

And what, friends, is right rapture?

Here individuals, aloof from sensual appetites, aloof from evil ideas, enter into and abide in true happiness, where there is, in addition to action and enjoyment, a time for meditation, which is born of solitude and is full of joy.

This, friends, is the truth concerning the way leading to the cessation of ill.

ENLIGHTENMENT AND IGNORANCE,
Islam

The predominance of an enlightened being over others is equal to the pre-eminence of the moon, at the night of the full moon, over all the stars.

Truly, the enlightened ones are the heirs of the prophets.

The placing of true knowledge before one who does not appreciate it is like placing a necklace of pearls, jewels and gold on the necks of swine.

Truly, the enlightened ones are the heirs of the prophets.

They die not who give themselves to learning about God. Religion is a store, and wisdom the route to it. The chief aim of true knowledge is virtue.

Truly, the enlightened ones are the heirs of the prophets.

Humility is the outcome of true knowledge. Minds are locked-up stores; only questions open them. Every look supplies a lesson to the wise. The works of those who have gone before are instructive to those who follow.

Truly, the enlightened ones are the heirs of the prophets.

One's following of the way of the Infinite is proportional to one's wisdom. To fight against one's own cravings is highest wisdom. A chief aim of wisdom is to know one's ignorance in relation to the whole.

Truly, the enlightened ones are the heirs of the prophets.

To separate oneself from things of time and to connect with things of eternity is highest wisdom. They are really wise whose actions attest to their words. Wise people are gainers in whatever condition they may be.

Truly, the enlightened ones are the heirs of the prophets.

The wise understand the ignorant, for they were once ignorant themselves. The wise aim at perfection; the foolish aim at wealth. People are often enemies of what they are ignorant of.

***Truly, the enlightened ones are the heirs of the prophet*s.**

EXCERPTS FROM JESUS' Sermon on the Mount

Blessed are the poised in spirit, for theirs is lasting happiness.

Blessed are the meek, for they shall inherit the earth.

Blessed are they that hunger and thirst after righteousness, for they shall be filled.

Blessed are the merciful, for they shall obtain mercy.

Blessed are the pure in heart, for they shall see God.

Blessed are the peacemakers, for they shall be called manifestations of the Infinite.

Blessed are those who are willing to suffer for righteousness' sake, for theirs is lasting happiness.

Blessed are you even if people should revile you and persecute you and utter all kinds of evil against you falsely.

Accept their rebukes and be calm, for people also persecuted the prophets who were before you.

You are the salt of the earth; but if salt has lost its taste, how can its saltiness be restored?

It is no longer good for anything except to be thrown out and trodden under foot.

We must work the works of the Eternal while it is day, for the night comes, when no one can work.

You are the light of the world.

A city set on a hill cannot be hid.

Nor do people light a lamp and put it under a bushel, but on a stand, and it gives light to all in the house.

Let your light shine, that all may see your good works.

Do not lay up for yourselves treasures because of material worth, but lay up for yourselves treasures of spiritual value.

For where your treasure is, there will be your heart be also.

Do not be anxious about tomorrow, for tomorrow will be anxious for itself.

Let the day's own trouble be sufficient for the day.

Do not judge others, that you may not be judged in return.

For with the judgment you pronounce you will be judged, and the measure you give will be the measure you get.

Ask, and it will be given you; seek, and you will find; knock, and it will be opened to you.

For everyone who asks, receives and the one who seeks finds;

And to all who knock it will be opened.

FAITH, HOPE AND LOVE, Paul the Apostle

If I speak with the language of the greatest of people, but have not love, I am a noisy gong or a clanging cymbal.

And if I have prophetic powers, and understand all mysteries and all knowledge, and if I have all faith, so as to remove mountains, but have not love, I am nothing.

If I give away all that I have, and if I deliver my body to be burned, but have not love, I gain nothing.

Love is patient and kind; love is not jealous or boastful; it is not arrogant or rude.

Love does not insist on its own way; it is not irritable or resentful; it does not rejoice at wrong, but rejoices in the right.

Love bears all things, believes all things, hopes all things, endures all things.

Love never ends; as for prophecy, it will pass away; as for languages, they will cease; as for knowledge, it will pass away.

For our knowledge is imperfect and our prophecy is imperfect: but when the perfect comes, the imperfect will pass away.

When I was a child, I spoke as a child, I thought as a child, I reasoned like a child: when I became an adult, I gave up childish ways.

For now we see in a mirror dimly, but then face to face.

Now I know in part; then I shall understand fully.

So faith, hope, love abide these three; but the greatest of these is love.

IF, Rudyard Kipling

If you can keep your head when all about you
Are losing theirs and blaming it on you:

*If you can trust yourself when everyone doubts you,
but make allowance for their doubting too:*

If you can wait and not be tired by waiting,
Or being lied about, don't deal in lies,

*Or being hated, don't give way to hating
and yet don't look too good, nor talk too wise;*

If you can dream and not make dreams your master;
If you can think and not make thoughts your aim:

*If you can meet with triumph and disaster
and treat those two impostors just the same:*

If you can bear to hear the truth you've spoken
Twisted by knaves to make a trap for fools,

*Or watch the things you gave your life to, broken,
and stoop and build 'em up with worn-out tools:*

If you can make one heap of all your winnings
And risk it on one turn of pitch-and-toss,

And lose, and start again at your beginnings
and never breathe a word about your loss:

If you can force your heart and nerve and sinew
To serve your turn long after they are gone,

And so hold on when there is nothing in you
except the will which says to them: "hold on!"

If you can talk with crowds and keep your virtue,
Or walk with rulers - nor lose the common touch;

If neither foes nor loving friends can hurt you,
if each one counts with you, but none too much:

If you can fill the unforgiving minute
With sixty seconds' worth of distance run,

Yours is the earth and everything that's in it,
and - which is more - you'll be an adult, my son
.. and my daughter, every one ..

MANY ASPECTS, ONE TRUTH, Ramakrishna

You see many stars at night in the sky but find them not when the sun rises; can you say there are no stars in the sky during the day?

So, O friend! Because you do not find the Eternal in the days of your ignorance, do not say there is no Eternal.

As one and the same material, water, is called by different names by different peoples; one calling it water, another eau, a third aqua, another pani;

So the One, the Everlasting-Intelligent-Bliss, is sought by some as God, by some Allah, by some as Jehovah, by some as Hari, and by others as Brahman.

As one can ascend to the top of a house by means of a ladder or a bamboo or a staircase or a rope,

So diverse are the ways and means of approaching the Eternal, and every religion in the world shows at least one of these ways.

Different creeds are but different paths to reach the Eternal.

Various and different are the ways that lead to the temple of Mother Kali.

Similarly, various are the ways that lead to the dwelling place of the Divine.

Every religion is nothing but one or more of such paths that lead to the Eternal.

As a young wife in a family shows her love and respect to every other member of the family, and at the same loves her husband more than these,

Similarly, being firm in your devotion to the supreme value of your own choice, do not despise other deities and beliefs.

But honor them all in proportion to the thoughts and actions they produce.

The One likewise has many forms.

The devotee who has seen the Eternal in one aspect only, knows it in that aspect alone.

But he who has seen it in manifold aspects is alone in a position to say, "All these forms are one, and the Eternal is mutltiform".

It is formless and with form, and many are its forms which no one knows.

ON CHILDREN, Kahlil Gibran

And a woman who held a babe against her bosom said, Speak to us of children. And he said,

Your children are not your children. They are the sons and daughters of Life's longing for itself.

They come through you but not from you, and though they are with you, yet they belong not to you.

You may give them your love but not your thoughts, for they have their own thoughts.

You may house their bodies but not their souls,

For their souls dwell in the house of tomorrow, which you cannot visit, not even in your dreams,

You may strive to be like them, but seek not to make them like you.

For life goes not backward nor tarries with yesterday.

You are the bows from which your children as living arrows are sent forth.

The Archer sees the mark upon the path of the Infinite, and bends you with such might that the arrows may go swift and far.

Let your bending in the Archer's hand be for gladness.

For even as the arrow that flies is beloved, so also is the bow that is stable.

ON DEATH, Kahlil Gibran

Then a woman spoke, saying, We would ask now of Death. And he answered,

You would know the secret of death. But how shall you find it unless you seek it in the heart of life?

The owl whose night-bound eyes are blind to the day cannot unveil the mystery of light. If you would indeed behold the spirit of death, open your heart wide to the body of life.

For life and death are one, even as the river and the sea are one.

In the depth of your hopes and desires lies your silent knowledge of the beyond;

And like seeds dreaming beneath the snow, your heart dreams of spring.

Trust the dreams, for in them is hidden the gate to eternity.

Your fear of death is but the trembling of the shepherds standing before the ruler whose hand is to be laid upon them in honor.

Are the shepherds not joyful beneath their trembling, that they shall wear the mark of the ruler?

Yet are they not more mindful of their trembling?

For what is it to die but to stand naked in the wind and to melt into the sun?

And what is it to cease breathing, but to free the breath from its restless tides,

That it may rise and expand and seek God unencumbered?

Only when you drink from the river of silence shall you indeed sing.

And when you have reached the mountain top, then you shall begin to climb.

And when the earth shall claim your limbs, then shall you truly dance.

ON REASON AND PASSION, Kahlil Gibran

And the priestess spoke and said: Speak to us of Reason and Passion. And he answered saying:

Your soul is oftentimes a battlefield, upon which your reason and your judgment wage war against your passion and your appetite.

Would that I could be the peacemaker in your soul, that I might turn the discord and the rivalry of your elements into oneness and melody.

But how shall I, unless you yourselves be also the peacemakers even the lovers of all your elements?

Your reason and your passion are the rudder and the sails of your seafaring soul.

If either your sails or your rudder be broken, you can but toss and drift, or else be held at a standstill in midseas.

For reason, ruling alone, is a force confining; and passion, unattended is a flame that burns to its own destruction.

Therefore let your soul exalt your reason to the height of passion, that it may sing;

And let it direct your passion with reason that your passion may live through its own daily resurrection, and like the phoenix rise above its own ashes.

I would have you consider your judgment and your appetite even as you would two loved guests in your house.

Surely you would not honor one guest above the other; for who is more mindful of one loses the love and the faith of both.

Among the hills, when you sit in the cool shade of the white poplars, sharing the peace and serenity of distant fields and meadows - then let your heart say in silence, "Eternity is to be found in reason".

And when the storm comes, and the mighty wind shakes the forest, and thunder and lightning proclaim the majesty of the sky - then let your heart say in awe, "Eternity is found also in passion".

And since you would be a breath in the eternal sphere, and a leaf in the everlasting forest, you too should rest in reason and move in passion.

ONENESS OF HUMANITY, Baha'i

This is a new cycle of human power.

All the horizons are luminous, and the world will indeed become as a garden and a paradise.

It is the hour of unity among all peoples, and of the drawing together of all races and all classes.

The gift to this age is the knowledge of the oneness of humanity and of the fundamental oneness of religion.

War shall cease, and the Most Great Peace shall come.

The world shall he seen as a new world, and all beings shall live as brothers and sisters.

Oneness means that the spirit of God should be realized as the power which animates and pervades all things, which are but the manifestations of its energy.

The tabernacle of unity is raised. Look not upon each other with the eye of strangeness!

Commit not that which will disturb the clarity of the pure water of love or sever the perfumed ties of friendship.

0 peoples of earth! Turn from the darkness of foreignness to the shining of the sun of unity;

This is that which shall benefit the people of the world more than anything else.

The tree of truth has no better blossom.

And the ocean of wisdom shall never have a brighter pearl than this.

The lights of the oneness of humanity are sparkling like precious jewels;

Scatter their rays to all parts!

Praise the banner of unity, fraternity, cooperation, and good relations among all the people,

So that there may not be left of prejudice anything but a name, and of ignorance not even a shadow;

That true religion may pitch its tent over the east and the west,

The light of God illumine all hearts,

Perfect understanding and association between races, religions and nationalities be realized.

That the world of creation may find security and peace!

THE ETERNAL, Upanishads

In the heart of all things, of whatever there is in the universe, dwells the Eternal. It above all is the reality.

Therefore, renouncing vain appearances, rejoice in it. Covet no person's wealth.

Well may they be to live a hundred years who act without craving - who work with earnestness, not yearning for its fruits - they, and they alone.

The Self is one. Unmoving, it moves swifter than thought.

The senses do not overtake it, for always it goes before.

Remaining still, it outstrips all that runs. Without the Self there is no life.

To the ignorant the Self appears to move - yet it moves not.

From the ignorant it is far distant - yet it is near.

It is within all, and it is without all.

Those who see all beings in the Self, and the Self in all beings, hate none.

To the illumined soul the Self is all.

For those who see everywhere oneness, how can there be delusion or grief?

The Self is everywhere. Bright is it, bodiless, without scar or imperfection, without bone without flesh, pure, untouched by evil.

The One which is above all and in all, the Eternal - this it is that has established order among objects and beings from beginningless time.

They truly know the Eternal who know it as beyond knowledge.

The ignorant think that the Eternal is known, but the wise know it to be beyond knowledge.

Those who realize the existence of the Eternal behind every activity of their being - whatever sensation, perception, or thought - they alone gain immortality.

Through knowledge of the Eternal comes power.

Through knowledge of the Eternal comes victory over death.

THE OLD AND THE NEW, Jesus the Christ

You have heard that it was said to the people of old, "You shall not kill".

But I say to you that everyone who remains angry with another is also at fault..

So if you are offering your gift to the church and there remember that someone has something against you, leave your gift there at the church and go: first be reconciled to that person, and then come and offer your gift.

Make peace with your adversary quickly, while you are still able to be a friend.

You have heard that it was said, "You shall not commit adultery".

But I say to you that everyone who looks at someone lustfully has already committed adultery within.

You have heard that it was said, "An eye for an eye, and a tooth for a tooth".

But I say to you, do not reject one who is evil; only resist evil itself.

If anyone strikes you on the right cheek, turn the other also:

And if people would sue you and take your coat, let them have your cloak as well;

And if anyone forces you to go one mile, go two miles.

Give to one who begs from you, and do not refuse one who would honestly borrow from you.

You have heard that it was said, "You shall love your neighbor and hate your enemy".

But I say to you, love your enemies, and do good to those who persecute you.

For if you love only those who love you, what merit have you shown?

And if you salute only your friends, what more are you doing than others?

THE WAY OF WISDOM, Hebrew Proverbs

That you may know wisdom and instruction,

Understand words of insight.

Receive instruction in wise dealing.

Righteousness, justice, and equity;

That prudence may be given to the simple,

Knowledge and discretion to the youth -

The wise also may hear and increase in learning,

And those of understanding acquire skill,

To understand a proverb and a figure,

The words of the wise and their riddles.

Hear, my child, your father's instruction,

And reject not your mother's teaching;

For they are a fair garland for your head,

And pendants for your neck.

My child, do not walk in the way with sinners,

Hold back your foot from their paths:

For their feet run to evil,

And they make haste to shed blood.

For in vain is a net spread

In the sight of any bird;

But these lie in wait for their own blood,

They set an ambush for their own lives.

Such are the ways of all who get gain by violence;

It takes away the life of its possessors.

Happy is the person who finds wisdom,

And those who get understanding,

For the gain from it is better than gain from silver

And its profit better than gold.

Wisdom is more precious than jewels,

And nothing you desire can compare.

Eternity is in wisdom's right hand:

In the left hand are true riches and honor.

Wisdom's ways are ways of happiness,

And all its paths are peace.

Wisdom is a tree of life to those who lay hold of it;

Those who hold it fast are called happy.

VARIETIES OF GIFTS, Paul the Apostle

Now concerning spiritual gifts, friends, I do not want you to be uninformed.

There are varieties of gifts, but the same spirit; and there are varieties of service, but the same ultimate goal.

And there are varieties of working, but it is the same spirit which inspires them all in everyone.

To each is given the manifestation of the spirit for the common good.

To one is given through the spirit the utterance of wisdom, and to another the utterance of knowledge according to the same spirit,

To another faith by the same spirit, to another prophecy,

To another the ability to distinguish between people, to another various kinds of languages, to another the interpretation of languages.

All these are inspired by one and the same spirit.

For just as the body is one and has many members, and all the members of the body, though many, are one body, so it is with humankind.

For the body does not consist of one member but of many.

If the foot should say, "Because I am not a hand, I do not belong to the body", that would not make it any less a part of the body.

If one member suffers, all suffer together; if one member is honored, all rejoice together.

Now you are the body of humanity and individually members of it.

And there are among you first religious leaders, second prophets, third teachers, then helpers, administrators, speakers in various kinds of languages.

Are all religious leaders? Are all prophets? Are all teachers? Do all speak with different languages? Do all interpret?

But earnestly desire the higher gifts.

WE ARE THE NEW CIVILIZATION.
Flemming Funch

We are here,

We are waking up now, out of the past, to dream a bigger dream.

We are friends and equals, we are diverse and unique, and we're united for something bigger than our differences.

We believe in freedom and cooperation, abundance and harmony.

We are a culture merging a renaissance of the essence of humanity.

We find our own guidance, and we discern our own truth.

We go in many directions, and yet we refuse to disperse.

We have many names, we speak many languages.

We are local, we are global.

We are in all regions of the world, we're everywhere in the air.

We are universe being aware of itself, we are the wave of evolution.

We are in every child's eyes, we face the unknown with wonder and excitement.

We're messengers from the future, living in the present.

We come from silence, and we speak our truth.

We cannot be quieted, because our voice is within everyone.

We have no enemies, no boundaries can hold us.

We respect the cycles and expressions of nature, because we are nature.

We don't play to win, we play to live and learn.

We act out of inspiration, love and integrity.

We explore, we discover, we feel, and we laugh.

We are building a world that works for everyone.

We endeavor to live our lives to their fullest potential.

We are independent, self-sufficient and responsible.

We relate to each other in peace, with compassion and respect, we unite in community.

We celebrate the wholeness within and around us all.

We dance to the rhythm of creation.

We weave the threads of the new times.

We are the new civilization.

SONGS

OF

INSPIRATION

AMAZING GRACE

Amazing grace! how sweet the sound,
That saved a soul like me!
I once was lost, but now am found,
Was blind, but now I see.
Twas grace that taught my heart to fear,
And grace my fears relieved;
How precious did that grace appear
The hour I first believed!
Through many dangers, toils and snares,
I have already come;
Tis grace has brought me safe thus far,
And grace will lead me home.
When we've been there ten thousand years,
Bright shining as the sun,
We've no less days to sing God's praise
Than when we first begun.

John Newton

AMAZING GRACE

American traditional
text: John Newton
arr.: D. Stone

A - maz - ing Grace, how sweet the
'Twas grace that taught my heart to
Through man - y dan - gers, toils and
When we've been here ten thou - sand

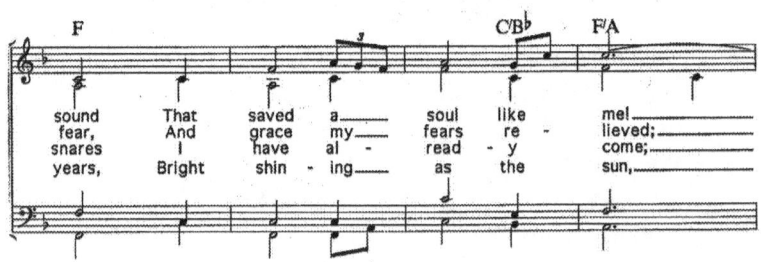

sound That saved a soul like me!
fear, And grace my fears re - lieved;
snares I have al - read - y come;
years, Bright shin - ing as the sun,

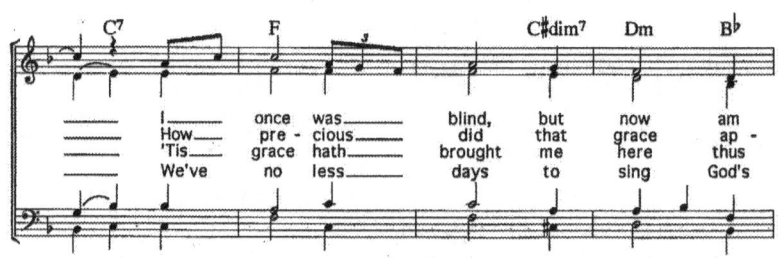

I once was blind, but now am
How pre - cious did that grace ap -
'Tis grace hath brought me here thus
We've no less days to sing God's

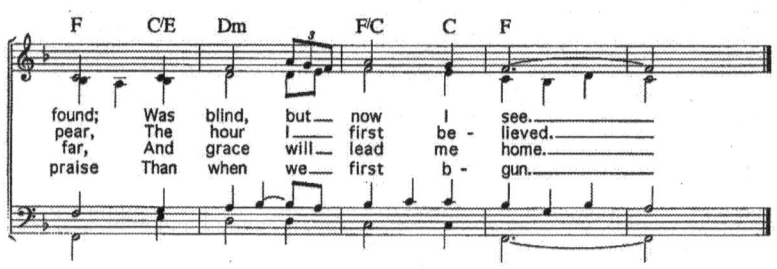

found; Was blind, but now I see.
pear, The hour I first be - lieved.
far, And grace will lead me home.
praise Than when we first b - gun.

(public domain)

AMEN

When we sing our hearts entwine
Amen! Amen!
In your face I see God shine.

Amen! Amen!
Amen, amen, amen,
Amen, amen,
Amen, amen, a men,
Amen, amen

When we dance my spirit soars..
Amen. Amen.
Learning, turning more and more.

Amen. Amen.
Amen, amen, amen,
Amen. Amen.
Amen, amen, amen.
Amen. Amen.

When I leave I hear Amen. Amen.
Sacred is this space and time.

Amen. Amen.
Amen, amen, amen.
Amen, amen,
Amen, amen, amen,
Amen, amen.

When we leave we keep in mind, Amen. Amen.
Sacred is this space and time. Amen, Amen.

Amen. Amen.
Amen, amen, amen,
Amen, amen.
Amen, amen, amen,
Amen, amen.

Richard Knox

AMEN

Richard Knox

AMEN

BRIDGES OF LOVE

We can build great bridges
Across the mighty waves between distant ridges.
Is it a task too great to build a bridge
Across the deaths of hate?
For now, more than ever,
What the world needs more of
Is to reach, to reach for each other,
With bridges of love.
If we can reach so far
To send men up to the moon and rockets to the stars,
Why are we still so far apart?
Why can't we find the way from soul to soul,
From heart to heart?
For now, more than ever,
What the world needs more of
Is to reach, to reach for each other,
With bridges of love.
Bridges of peace reach from shore to shore;
Bridges of love reach so much more.
They link our common hopes, our common ground,
Joining one and all the whole world 'round.
We all can build bridges of love each day
With our eyes, our smiles, our touch,
Our will to find a way.
There is no distance we cannot span;
Oh, the vision is in our hearts,
The power is in our hands.
For now, now more than ever,
What the world needs more of
Is to reach, to reach for each other,
With bridges of love.
Oh, oh, now, now more than ever,
What the world needs more of
Is to reach, to reach for each other,
With bridges of love.

Stephen Longfellow Fiske

BRIDGES OF LOVE

Stepehen Longfellow Fiske

If we can build great bridg-es a-cross the might-y waves be-tween dis-tant ridg-es is it a task too____ great____ to build__ a bridge a-cross the depths of hate?____ For now more than ev-er, what the world needs more of____ is to reach, to reach for each oth-er____ with bridg-es of love. If we can__ reach so far____ to send men up to the moon, and rock-ets to the stars, why are we still so far__ a-part? Why can't we find the way from soul to soul, from

— 1 —

BRIDGES OF LOVE

BRIDGES OF LOVE

will to find a way. There is no dis - tance____ we can - not span Whoa,_ the

vi - sion is in our hearts, the pow - er is in our hands.____ For.

now, now more than ev - er,____ what the world needs more

of____ is to reach to reach for each oth- er__ with bridg - es of

love. Whoa!____

EARTH ANTHEM

O say, can we see
By the one light in all—
Our earth to embrace
At the call of all nations;
Where our children can play
In a world without war,
Where we stand hand in hand
In the grace of creation;
Where the rivers run clean
Through the forests of green;
Where the cities stand tall
In the clear skies of freedom
O say, do our hearts sing
For harmony and love forever—
On the planet of our birth,
Blessed with peace on earth!

Words: Stephen Longfellow Fiske 310-396-8205

EARTH ANTHEM

Lyrics: Stephen Longfellow Fiske
Music: John Stafford Smith, ca 1771

EARTH ANTHEM

GATHER US IN, O LOVE

Gather us in, O Love, that fills us all;
Gather our rival faiths within your fold;
Rend each one's temple veil, and bid it fall,
That we may know that you have been of old.

Gather us in, we worship only you;
In varied names we stretch a common hand;
In diverse forms a common soul we view;
In many ways we seek one promised land.

Yours is the mystic life great India craves;
Yours is the Parsee's sin-destroying beam;
Yours is the Buddhist's rest from tossing waves;
Yours is the empire of vast China's dream.

Yours is the Roman's strength without his pride:
Yours is the Greek's glad world without its graves:
Yours is Judea's law with love beside,
The truth that censures and the grace that saves.

Some seek a Father in the heav'ns above:
Some ask a human image to adore;
Some crave a spirit vast as life and love:
Within your mansions we have all and more.

George Matheson

GATHER US IN, O LOVE

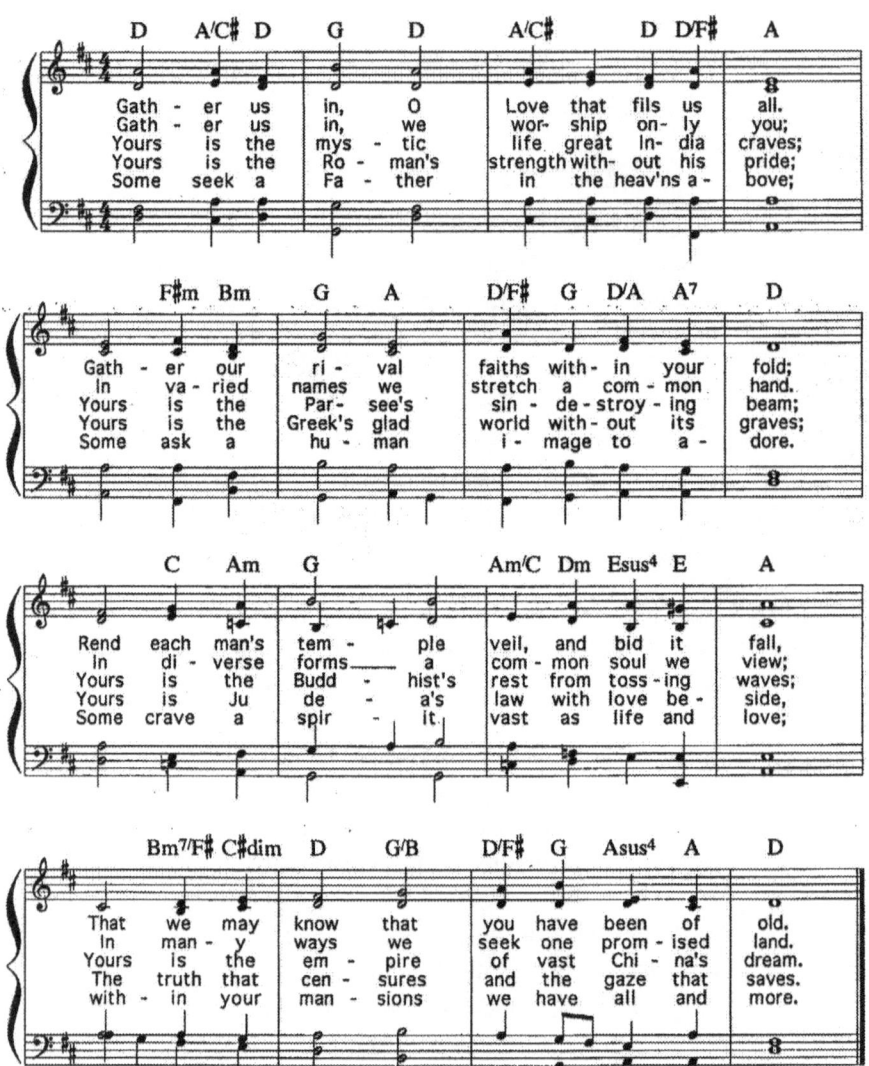

HEAR, HEAR, O YOU NATIONS

Hear, hear, o you nations, and hearing obey
The cry from the past and the call of today!
Earth wearies and wastes with her fresh life outpoured,
The glut of the cannon, the spoil of the sword.

Lo, dawns a new era, transcending the old,
The poet's rapt vision, by prophet foretold!
From war's grim tradition it makes its appeal
To service of all in a world's commonweal.

And you, O my country, from many made one,
Late born among nations, at morning your sun,
Arise to the place you are given to fill,
And lead the world-triumph of peace and goodwill.

Frederick Lucine Hosmer

HEAR, HEAR, O YOU NATIONS

HIGH UPON A MOUNTAIN

High, high upon a mountain I have a secret shrine.
High, high above the stormy clouds, there the sun will always shine.

I don't follow a winding highway and it's not on any map I've ever known.
In my heart there's a special sky-way that leads to a place all my own.

High, high upon a mountain I go alone to pray.
All nature holds me in its arms, all my burdens fall away.

I don't follow a winding highway and it's not on any map I've ever known,
In my heart there's a special sky-way that leads to a place all my own.

High, high upon a mountain I have a secret shrine.
All heaven sings inside my heart. Peace and happiness are mine.

Jan-Lee Music 1955 - 1983
Used by permission

HIGH UPON A MOUNTAIN

HIGH UPON A MOUNTAIN

fol - low_____ a wind - ing high - way, and it's not on an - y map I've ev - er

Cm⁷   F⁷   Bb⁷   Bbm⁷   Eb⁷

known.   In   my   heart there's a spe - cial sky - way   that

Abmaj⁷   Cm⁷   F⁷   Gm⁷   Bb/F

leads to a place all my own.   mine.

Cm⁷   F⁹   Bb⁷sus⁴   Bb⁷   Eb   Eb⁶   Ab⁶   Eb⁶

— 2 —

239

THE IMPOSSIBLE DREAM

To dream the impossible dream,
To fight the unbeatable foe,
To bear with unbearable sorrow,
To run where the brave dare not go;

To right the unrightable wrong,
To love, pure and chaste from above,
To try when your arms are too weary,
To reach the unreachable star -

This is my quest, to follow that star,
No matter how hopeless, no matter how far.
To fight for the right, without question or pause;
To be willing to march into hell for a heavenly cause;

And I know if I'll only be true to this glorious quest,
That my heart will lie peaceful and calm
When I'm laid to my rest;
And the world will be better for this,
That one man scorned and covered with scars,
Still strove with his last ounce of courage
To fight the unbeatable foe,
To reach the unreachable star!

From "Man of La Mancha"

IT'S A SMALL WORLD

It's a world of laughter, a world of tears
It's a world of hopes and a world of fears,

There's so much we share
That it's time we're aware
It's a small world after all.

It's a small world after all,
It's a small world after all,
It's a small world after all,
It's a small, small world.

There is just one moon and one golden sun
And a smile means friendship to everyone
Though the mountains divide and the oceans are wide
It's a small world after all.

It's a small world after all,
It's a small world after all,
It's a small world after all,
It's a small, small world .

KUM BAH YAH

Kum bah yah, my Lord, kum bah yah,

Kum bah yah, my Lord, kum bah yah,

Kum bah yah, my Lord, kum bah yah.

Oh, Lord, Kum bah yah.

2.   Someone's singing,   Lord, kum bah yah,

3.   Someone's mourning   Lord, kum bah yah,

4.. Someone's praying,   Lord, kum bah yah.

Oh, Lord, Kum bah yah.

LET THERE BE PEACE ON EARTH

Let there be peace on earth
And let it begin with me
Let there be peace on earth
The peace that was meant to be.

With God as our creator *
Children all are we
Let us walk with each other
In perfect harmony
Let peace begin with me
Let this be the moment now
With every step I take
Let this be my solemn vow
To take each moment
And live each moment in peace eternally
Let there be peace on earth
And let it begin with me.

Sy Miller & Jill Jackson Copyright 1955 Jan-Lee Music Copyright renewal 1983 Used by permission
* Original version— "With God as our Father, Brothers all are we; Let me walk with my brother"

LET THERE BE PEACE ON EARTH

LET THERE BE PEACE ON EARTH

LIGHT THE TORCH

Light the torch
Prepare the way
Start the parade
This is the day
We've come together
From near and far
To reach our highest
To be all that we are.

We join together
To set a new pace
For all the children
Of the human race
We may earn a medal
As we win or score
But what we are doing
Is worth even more

Light the torch
Prepare the way
Start the parade
This is the day
We've come together
From near and far
To reach our highest
To be all that we are.

Over the hurdles
As we set a new mark,
We challenge the future
We light a spark
To make the goals higher
Higher and higher
To be our best
Our one desire

As we feel the unity
Of bodies and minds
Working together
Our spirit we find
The spirit of harmony
And brotherhood
Of peace and love
Of all that's good.

Light the Torch
Prepare the way
Start the parade
This is the day
We've come together
From near and far
To reach our highest
To be all that we are

Light the Torch

Lyrics: Breeze Bryson, Corinee Schriner
Music: Breeze Bryson, Rick Lewis, Denise Elliott
Copyright: Touch & See, 1983

LIGHT THE TORCH

Breeze Bryson et al.

LIGHT THE TORCH

LIGHT THE TORCH

good. Light the torch;— Pre-pare the way! Start the pa- rade;— This is the day! We've come to-geth-er from near and far To reach our high-est To be all that we are.— Light the torch!

MAKE CHANNELS FOR THE STREAMS OF LOVE

Make channels for the streams of love,
Where they may broadly run
And love has overflowing streams
To fill them every one.

But if at any time we cease
Such channels to provide,
The very fount of love for us
Will soon be parched and dried.

For we must share if we would keep
That blessing from above -
Ceasing to give, we cease to have:
Such is the law of love.

Richard Chenevix Trench

MAKE CHANNELS FOR THE STREAMS OF LOVE

Richard Chenevin Trench

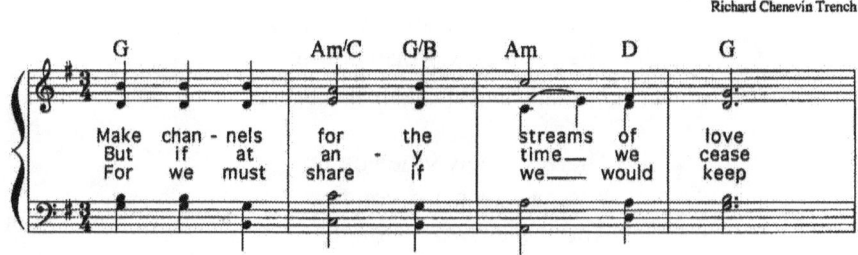

Make chan - nels for the streams of love
But if at an - y time__ we
For we must share if we__ would

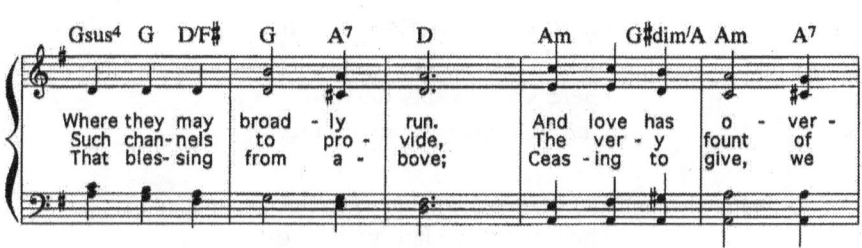

Where they may broad - ly run.
Such chan - nels to pro - vide,
That bles- sing from a - bove;
And love has o - ver -
The ver - y fount of
Ceas - ing to give, we

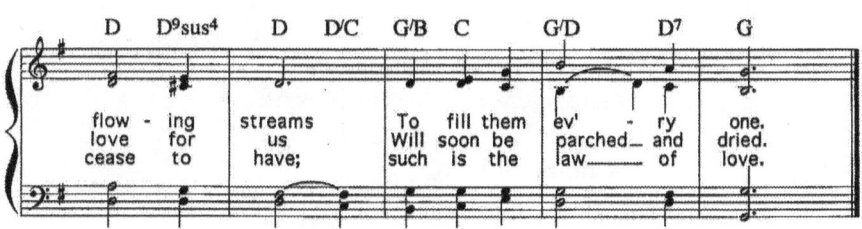

flow - ing streams To fill them ev' - ry one.
love for us Will soon be parched_ and dried.
cease to have; such is the law__ of love.

251

NEW WORLD COMIN'

There's a new world comin',
And it's just around the bend;
There's a new world comin',
This one's comin' to an end.

There's a new voice callin'
You can hear it if you try.
And it's growin' stronger
With each day that passes by.

There's a brand new mornin'
Risin' clear and sweet and free;
There's a new day dawnin'
That belongs to you and me.

Yes, a new world's comin',
The one we've had visions of;
Comin' in peace, comin' in joy,
Comin' in love.

Yes, a new world's comin'
The one we've had visions of;
Comin' in peace, comin' in joy,
Comin' in love.

Comin' in love, comin' in love....

NO ONE IS AN ISLAND

No one is an island,
No one stands a-lone,
Each one's joy is a joy to me,
Each one's grief is my own.
We need one another,
Or so I will de-fend
Each one is my sibling.
Each one is my friend.

I saw the people gather,
I heard the mu-sic start,
The song that they were singing
Is ringing in my heart.

No one is an island,
No one stands a-lone,
Each one's joy is a joy to me,
Each one's grief is my own.
We need one another,
Or so I will de-fend
Each one is my sibling,
Each one is my friend.

NO ONE IS AN ISLAND

NO ONE IS AN ISLAND

NOT IN VAIN THE DISTANT BEACONS

Not in vain the distant beacons,
Forward, forward let us range,
Let the great world spin forever
Down the ringing grooves of change;
Thru the shadow of the globe we
Sweep ahead to heights sublime,
We, the heirs of all the ages
In the foremost files of time.

O, we see the crescent promise
Of our spirit has not set;
Ancient founts of inspiration
Well thru all our fancy yet;
And we doubt not through the ages
One increasing purpose runs,
And the thoughts of all are widened
With the process of the suns.

Yes, we dip into the future,
Far as human eye can see,
See the vision of the world, and
All the wonder that shall be,
Hear the war drum throb no longer,
See the battle flags all furled,
In the Parliament of all, the
Federation of the World.

NOT IN VAIN THE DISTANT BEACONS

Ludwig van Beethoven
(adapted from the 9th Symphony)

Not in vain the dis- tant beac-cons For- ward, for- ward let us range,
O, we see the cres- cent prom- ise Of our spir - it has not set;
Yes, we dip in- to the fu- ture, Far as hu- man eye can see,

Let the great world spin for- ev - er Down the ring-ing grooves of change;
An-cient founts of in- spir- a- tion Well through all our fan - cy yet;
See the vi - sion of the world, and All the won-der that shall be.

Through the shad- ow of the globe we Sweep a-head to heights sub- lime, We
And we doubt not through the a-ges One in-creas- ing pur- pose runs. And
Hear the war drum throb no long- er See the bat-tle flags all furled, In

— the heirs of all the a- ges In the fore-most files of time.
— the thoughts of all are wid- ened With the pro- cess of the suns.
— the Par - lia- ment of all, the Fed-er - a - tion of the World.

ONE GREAT MOMENT

If we want a world filled with peace,
a world where all people are fed,
then we'll make a world safe for everyone -
white, black, yellow, brown and red.

If we can have peace for a minute,
then we can have peace for an hour
If we can have peace all together
then we can have peace forever.

One great moment of peace!
One great moment when we all will cease
all our conflicts, all our fears,
all our wars and all our tears,
One great moment of peace!

One great moment of joy!
for every girl and every boy,
every woman, every man,
every country, every land,
One great moment of joy!

One great moment of love!
In the spirit of the dove,
in the spirit of unity,
joining hearts from sea to sea.
One great moment of love!

One great moment in time!
One moment when we all will find
peace on earth and peace of mind,
love and joy for humankind.
One great moment in time!

ONE GREAT MOMENT FOREVER!

Words and music: Breeze Bryson. Copyright: Touch &. See, 1987

PEACEMAKERS

The gentle people are standing
On a hill above the city.
The new sun is rising
On the far horizon.
Hand in hand they stand;
The silence speaks its wisdom;
Drawing deep upon the well
Of peace that dwells within them.
Peacemakers, gather together;
Peacemakers, our hope is in you.
A hymn rose out among them
Like a breeze upon the water;
Reaching for the restless shores
Of the hearts of sons and daughters.
Working hard and struggling
Just to find a better way -
To live a life in dignity
With loved ones each day.
Peacemakers, gather together;
Peacemakers, our hope is in you.
We can all be peacemakers
If each day we live the belief
That within ourselves and around the world
We hold the key to peace.
And you may not be an ambassador,
Or a diplomat or a king;
You may just be a humble soul,
With a peaceful gift to bring.
Peacemakers, gather together;
Peacemakers, our hope is in you.
Peacemakers, gather together;
Peacemakers, our hope is in you.
Peacemakers, our hope is in you.
We are the peacemakers.
Gather together.
We're all one family under the sun.
Gather together, gather together.
We are the peacemakers;
Let the world hear our song.

Our hope is in you.
Let the world know we're waging peace.

Make peace, make peace, make peace, make peace, make peace.

Stephen Longfellow Fiske

RESTING IN GOD

Whatever I already am, I want to devote to life.
I want, deliberately, for life to make use of the best I have and
am.
I may not be sure at this moment in what way this could be,
even if I have
Ideas, I will allow for the greater intelligence and wisdom,
deep within me,
To guide me.

Now that we're sure that your love is never ending
And the choices we made here a-lone are not worth
defending,
Wake to our only protection
Then everywhere that we are
We are your reflection.
Nothing can threaten or harm us, we're safe in your arms. Ah
Content in our hearts and our mind, resting in God,
Resting in God, resting in God. Ho

Nothing can threaten or harm us, save in your arms...mm
Content in our hearts and our mind, resting in God.
Resting in God, Resting in God

Richard Knox

RESTING IN GOD

Richard Knox

Whatever I already am, I want to devote to life.
I want, deliberately, for life to make use of the best I have and am.
I may not be sure at this moment in what way this could be, even if I have
ideas, I will allow for the greater intelligence and wisdom, deep within me,
to guide me.

Mmm___

Now that we're sure that your love___ is nev - er - end - ing,___

And the choic - es we made here a - lone are not worth de - fend - in

In re - mem - ber - ing__ we a -

wake__ to our on - ly pro tec - tion___

Then ev' - ry - where that we are_ we are your re - flec - tion.___

RESTING IN GOD

No- thing can threat - en or harm us, (we're) safe in your arms. ah—

Con - tent in our hearts and our mind, rest - ing in

God. rest - ing in

God. rest - ing in God. Ho—

No- thing can threat - en or harm us, safe in your arms. mm.

Con - tent in our hearts and our mind, rest - ing—

in God, rest - ing in

God, rest - ing in God.

And all is well.

— 2 —

SAVE OUR WORLD

What are we doing and where are we going?
Why are we destroying our world?

Neighbors are dying...and
Children are crying.
Why can't we have peace in this world?

We should be striving,
To make life worth living
Sharing our bounties
Throughout this world.

Open your heart up;
We'll never give up
Let us all work hard....to
Save our world.

What are we doing and where are we going?
Why are we destroying our world?

Words and music by: Natalie LaCroix Forrest aka Rev. Nathalie LaCroix
Forrest, D.D.
A Gift to Unity-and-Diversity

SAVE OUR WORLD

Natalie LaCroix Forrest

SONG OF THE AVATARS

When righteousness declines
And wickedness is strong,
In the dying of an age
As a new age comes along,
That is when I rise again
And yet again to light the flame
Of truth within the hearts of all.

I am Light
I am Truth
I am the fire of the sun
I am the hope of all the earth,
The spirit of the One.

When all have lost their way
And know not where to turn.
And the future seems to end
Where the fires of hatred burn,
Then let them look within
To see the rising of my flame
Of love to light the way again.

I am Love
I am Truth
I am the power of the sun
I am the light within all beings,
The spirit of the One.

When darkness seeks to hold
The hearts of all in fear,
When all cry out for help
And no one seems to hear.
That is when I rise again
To shatter forms enslaving all
To let a newer world be born.

When righteousness declines
And wickedness is strong,
In the dying of an age

As a new age comes along
That is when I rise again
And yet again to light the flame
Of truth within the hearts of all.

I am Light
I am Truth
I am the freedom of the sun
I am the destiny of all,
The triumph of the One.

I am Light
I am Truth
I am the fire of the sun
I am the hope of all the earth,
The spirit of the One.

Words by David Spangler,
Music by Milenko Matanovic

SONG OF THE AVATARS

Lyric: David Spangler
Music: Milenko Matanovic

When right-eous-ness de-clines And wick-ed-ness is strong, In the dy-ing of an age as a new age comes a-long, That is when I rise a-gain and yet a-gain to light the flame Of truth with-in the hearts of all. I am Light. I am Truth.— I am the fire of the sun. I am the hope of all the earth, the spir-it of the One.

When all have lost their way And know not where to turn And the fu-ture seems to end where the fires of hat-red burn, Then let all look with-in To see the ris-ing fo my flame Of love to light the way a-gain: I am

SONG OF THE AVATARS

Love.___ I am Truth.___ I am the pow-er of the sun. I am the
light with - in us all, The spir - it of the One.

Verse 3
When dark-ness seeks to hold The hearts of all in fear, When
all cry out for help, And no one seems to hear, That is when I rise a-gain To shat-ter
forms en-slav-ing all To let a new - er world be born.

da Capo al Coda

Interlude

CODA
Light. I am Truth.___ I am the free - dom of the sun.
Light I am Truth.___ I am the fire___ of the sun.

I am the des - ti - ny of all, The tri - umph of the
I am the hope of all the earth, The spir - it of the

1.
2. G
One.___ I am ___

— 2 —

269

THE NEW AGE IS NOW

The New Age is here, The new age is now
At center's the way. In love is the now.

So, I'll let it begin simply to be,
I'll let love begin here, now with me.
The New Age begins here, now with me.

The seed forms are many, their energy one
Just tending the gardens, The New Age begun

So, I'll let it begin simply to be,
I'll let love begin here, now with me.
The New Age begins here, now with me.

Words and music C. 1976 Patricia James

THE NEW AGE IS NOW

Patricia Jones

The New Age is here_____ The New Age is now. At
The seed forms are ma - ny, their en - er - gy one. Just

cen - ter's the way._____ In love is the how. So I'll
tend - ing the gar - - den's the New Age be - gun So I'll

let it be - gin_____ sim - ply to_____ be. I'll

let love be - gin_____ here, now with me. The

New Age be - gins_____ here,__ now with me.

UNITY-'N-DIVERSITY

Greetings to you who serve the purpose
Who strive and work to set us free
Peace to all whose understanding spans
    Unity-'n-Diversity.

Love to all Brothers and Sisters
Love to all the Creator's Sons
To you who seek to learn the purpose
The Law of Love, the Law is One.
    Unity-'n-Diversity.

Joy is yours in the giving
Wisdom comes in times yet to be
Light is yours from the beginning of
    Unity-'n-Diversity

Love to all Brothers and Sisters
Love to all the Creator's Sons
To you who seek to learn the purpose
The Law of Love, the Law of One.
    Unity-'n-Diversity.

Live the Spirit and in so living
The life abundant you will see
Fret not o're what tomorrow brings you.
    Unity-'n-Diversity.

Love to all Brothers and Sisters
Love to all the Creator's Sons
To you who seek to learn the purpose
The Law of Love, the Law is One.
    Unity-'n-Diversity
    Unity-'n-Diversity
    Unity-'n-Diversity

Words and Music by Michelina Foster.
A Gift to Unity-and-Diversity

UNITY-'N-DIVERSITY

Michelina Foster

UNITY-'N-DIVERSITY

WE ARE THE WORLD

There comes a time when we heed a certain call
When the world must come together as one.
There are people dying,
And it's time to lend a hand to life,
The greatest gift of all.

We can't go on pretending day by day
That someone, somewhere will soon make a change,
We are all a part of God's great big family;
And the truth, you know,
Love is all we need.

We are the world; we are the children.
We are the ones who make a brighter day,
So let's start giving.
There's a choice we're making:
We're saving our own lives.
It's true and we'll make a better day -
Just you and me.

Send them your heart, so they'll know that someone cares,
And their lives will be stronger and free,
As God has shown us by turning stones to bread:
So we all must lend a helping hand.

Chorus

When you're down and out, there seems no hope at all.
But if you just believe there's no way we can fall.
Let us realize that a change can only come
When we stand together as one.

Chorus

WE SHALL OVERCOME

1. We shall overcome, we shall overcome,
   We shall overcome today.
   Oh, deep in my heart, I do believe
   We shall overcome today.

2. We shall live in peace,
3. We'll walk hand in hand
4. Black and white together, black and white together, now.
5. We are not afraid, we are not afraid today.
6. We shall overcome.

WHERE HAVE ALL THE FLOWERS GONE?

Where have all the flowers gone?
Long time passing.
Where have all the flowers gone?
Long time ago.
Where have all the flowers gone?
Young girls picked them every one.
When will they ever learn?
When will they ever learn?

Where have all the young girls gone?
Long time passing.
Where have all the young girls gone?
Long time ago.
Where have all the young girls gone?
Gone to young men every one.
When will they ever learn?
When will they ever learn?

Where have all the young men gone?
Long time passing.
Where have all the young men gone?
Long time ago.
Where have all the young men gone?
Gone to grave-yards every one.
When will they ever learn?
When will they ever learn?

Where have all the grave-yards gone?
Long time passing.
Where have all the grave-yards gone?
Long time ago.
Where have all the grave-yards gone?
Gone to flowers every one.
When will they ever learn?
When will they ever learn?

Pete Seeger
Copyright 1961 (renewed) by SANGA MUSIC, INC.

WHERE HAVE ALL THE FLOWERS GONE?

Pete Seeger/Joe Hickerson

INTERFAITH

CELEBRATION

GUIDE

An Interfaith Celebration is a gathering of people from many faiths, both ancient and modern, to meditate and pray, share the music and the messages of the different faiths, and to engage in dialogue for mutual understanding. Often these experiences also lead to service through the cooperation of the faiths involved.

An Interfaith Celebration Guide is a booklet or book which contains materials to enrich the interfaith celebration. This book **Science and Spirituality** contains these materials, and they are presented in a way that will also stimulate the more general reader to enjoy them, even though they are not going to be used in an interfaith celebration.

The specific materials in **Science and Spirituality** which can be used in the interfaith celebration are the *Declaration of Interdependence*, the *Meditation and Candlelighting* readings, the *Responsive Readings*, and the *Songs of Inspiration* (see Table of Contents).

The suggested Order of Celebration is as follows:
>Opening Music
>Attunement (opening meditation and prayer)
>Declaration of Interdependence
>Song of Inspiration
>Welcome and remarks by person hosting the meeting
>Candlelighting
>Calls to prayer and meditation from various faiths, followed by a silent period
>Love offering with musical background
>Messages from various faiths on the theme of the program
>Music
>Dialogue
>Closing song - *Let there be Peace on Earth*

ELEMENTS OF THE CELEBRATION

THEME

The theme is determined ahead of the meeting. A year's theme is suggested, with monthly themes to be related to the annual one.

OPENING MUSIC

The Celebration should begin with live music when possible, or if not, with a quiet form of recorded music to set the tone for the event.

ATTUNEMENT

This is the prayer and/or meditation at the beginning of the Interfaith Celebration to link all participants with the spiritual purpose of the celebration. It consists of a joining of hands in a circle, a few words of invocation, followed by a brief period of silence. It concludes with a squeezing of the hands, "Peace be with us all", or some other way of ending that period.

DECLARATION OF INTERDEPENDENCE

The Declaration of Interdependence or some relevant affirmation of the meaning of the Celebration, should be read by the group following the attunement.

SONG OF INSPIRATION

Following the Declaration, it is appropriate to have a song, led by a live musician if possible, to lift the spirits of everyone and tune into the theme of the Celebration.

WELCOME AND REMARKS

If the Celebration is held in a religious or spiritual center, then the host is generally invited to welcome the group and give a message related to the faith of that center. If it is held in the same place each time, then someone needs to welcome the group and focus on the theme.

CANDLELIGHTING

A selected person chooses one of the candlelighting subjects that best relates to the theme of the day. The following italicized words are read by the facilitator:

The lighting of the candles is a ritual designed to express unity-and-diversity in a beautiful and dramatic way. We have on the table in front of us a single lighted white candle, surrounded by six or more colored candles.

The ceremony begins by different people, one at a time, choosing quotes from the subject used for the candlelighting, then taking a match, getting it lit from the central white candle, and lighting one of the colored candles. When all of the candles have been lit, the facilitator says:

We have separated the white color into the colors of the rainbow. The white candle symbolizes unity and the various colored candles, diversity.

The ceremony closes with the words repeated three times by the participants:

SPIRIT IS ONE, PATHS ARE MANY!

CALLS TO MEDITATION AND PRAYER

A facilitator introduces this ritual and introduces the various faith representatives who will take part all at once, explaining that each call to meditation and/or prayer should be one or two minutes, and that it should flow from one faith to the next without interruption. After the last faith representative has spoken, there should be a brief silent time to include everyone and to allow time for in-depth attunement. The ritual should be ended with "Peace be with us all" or some similar closing.

LOVE OFFERING

One of the facilitators calls for the offering, which can be initiated with the following group reading or spiritual appeal similar to the following:

May this money we are about to give be blessed. It is a symbol of divine substance and energy. May it be redeemed from any impure influence, any attachment or craving. We appreciate it and give thanks for it; we will use it for good, right, and appropriate purposes.

MESSAGES

The brief messages are from previously chosen faith representatives and should be spoken after each person is introduced and should be about three minutes in length, depending upon the number of representatives.

MUSIC

At various places during the celebration, live musicians should be given time to play and/or sing. When possible, these songs should include group participation, but solos and group performances are also appropriate if related to the theme. In the absence of live musicians, recorded music can be used.

DIALOGUE

It is important to allow time during each Interfaith Celebration for an informal sharing of insights and/or asking of questions. Greater clarity and a feeling of completion usually results from having dialogue time before the closing.

CLOSING CANDLELIGHTING RITUAL

The candles are picked up by some of the participants, who then turn 180 degrees, facing away from the circle, symbolizing outreach to the world. The facilitator explains that part of the ceremony as follows:

Please take a candle, turn to face away from the circle and silently visualize our service to the world, then blow out the candle and return it to the table.

CLOSING SONG

The Celebration needs some form of closure. Generally, this can come through the singing of *Let There Be Peace on Earth* or a similar song. It sometimes can best be done by providing a time for a benediction or a series of sharings affirming the meaning of the Celebration to the individuals involved.

DEFINITIONS

The English language allows for whole ranges of thoughts, feelings, ideas, and emotions. It has the ability to incorporate words from other languages. It is a living language and the Oxford English Dictionary faithfully tries to keep up to date with usage. Language however still follows the human mind, and sometimes new expressions are forged and new terminology used. People may or may not have a common acceptance of the meaning of a term, as millions of people world-wide have acquired English as their second language. This list is included to assist readers in the shaping of their own understanding of terminology used by others which may be critical to the progress and development of our emerging civilization.

Oxford English Dictionary.

ART:

a. Human skill as an agent, human workmanship. Opposed to *nature*.

b. The application of skill to subjects of taste, as poetry, music, dancing, the drama, oratory, literary composition, and the like; esp. in mod. use: Skill displaying itself in perfection of workmanship, perfection of execution as an object in itself.

ARTIST:

a. One skilled in the 'liberal' or learned arts.

b *gen.* One who cultivates one of the fine arts, in which the object is mainly to gratify the aesthetic emotions by perfection of execution, whether in creation or representation.

It formerly included all who cultivated any of the *arts* presided over by the *Muses*, i.e. history, poetry, comedy, tragedy, music, dancing, astronomy; hence the application to actors, musicians, dancers, and perhaps Milton's 'artist' = *astronomer*.(obs)

CAPITAL:

a. Of or pertaining to the original funds of a trader, company, or corporation; principal; hence, serving as a basis for financial and other operations.

b. A capital stock or fund. i. Commerce. The stock of a company, corporation, or individual with which they enter into business and on which profits or dividends are calculated; in a joint-stock company, it consists of the total sum of the contributions of the shareholders. ii. Pol. Econ. The accumulated wealth of an individual, company or community, used as a fund for carrying on fresh production; wealth in any form used to help in producing more wealth. iii. Fixed Capital: that which remains in the owner's possession, as working cattle, tools, machinery, etc. Circulating, floating capital: that which is constantly changing hands or passing from one form into another, as goods, money etc.

CAPITALISM: The condition of possessing capital; the position of a capitalist; a system which favors the existence of capitalists.

CAPITALIST: One who has accumulated capital; one who has capital available for employment in financial or industrial enterprises.

CIVIL: a. Pertaining to citizens, their private rights, etc., hence relating to the body of citizens or commonwealth, political, public; also, pertaining to the citizen as distinct from the soldier; and citizen-like, polite, courteous, urbane.
b. Of or belonging to citizens; consisting of citizens, or people dwelling together in a community, as in *civil society*.
c. Of or pertaining to the whole body or community of citizens; pertaining to the organization and internal affairs of the body politic, or state.
d. Having proper public or social order; well-ordered, orderly, well-governed.
e In that social condition which accompanies and is involved in citizenship or life in communities; not barbarous: civilized, advanced in the arts of life.

CIVILIZATION: a. The action or process of civilizing or of being civilized.

b. Civilized condition or state; a developed or advanced state of human society; a particular stage or a particular type of this.

CIVILIZE: a. To make civil; to bring out of a state of barbarism, to instruct in the arts of life, and thus elevate in the scale of humanity; to enlighten, refine, and polish.
b. To make 'civil' or moral; to subject to the law of civil or social propriety.

COALESCE: a. To grow together or into one body.
b. To unite or come together, so as to form one.

COALESCENCE: a. The growing together of separate parts.
b. The combination or uniting into a single body.

COMMERCE: a. Exchange between people of the products of nature or art; buying and selling together; trading; exchange of merchandise, esp. as conducted on a large scale between different countries or districts; including the whole of the transactions, arrangements, etc., therein involved.
b. Intercourse in the affairs of life; dealings.

CONSCIENCE: a. Inward knowledge, consciousness; inmost thought, mind.
b The internal acknowledgment or recognition of the moral quality of one's motives and actions; the sense of right and wrong as regards things for which one is responsible; the faculty or principle which pronounces upon the moral quality of one's actions or motives, approving the right and condemning the wrong.
c. *Good conscience*: An approving conscience; a consciousness that one's acts, or one's moral state, are right.

CONSCIOUSNESS: a. Internal knowledge or conviction; knowledge as to which one has the testimony within oneself; exp. of one's own innocence, guilt, deficiencies, etc .

b. The state or fact of being mentally conscious or aware of anything.

c. The recognition by the thinking subject of its own acts or affections.

d.The totality of the impressions, thoughts, and feelings, which make up a person's conscious being.

CONSILIENCE: The fact of 'jumping together' or agreeing; coincidence, concurrence; said of the accordance of two or more inductions drawn from different groups of phenomena.

COSMOS: The world or universe as an ordered and harmonious system.

DEMOCRACY: a. Government by the people; that form of government in which the sovereign power resides in the people as a whole, and is exercised either directly by them (as in the small republics of antiquity) or by officers elected by them. In mod. use often more vaguely denoting a social state in which all have equal rights, without hereditary or arbitrary differences of rank or privilege.

b. A state or community in which the government is vested in the people as a whole.

DEMOCRATIC: Of the nature of, or characterized by, democracy; advocating for upholding democracy.

DIRECTIVE: Having the quality or function of directing authoritatively guiding, or ruling.

ECONOMY: a. Management of a house; management generally. The art or science of managing a household, *esp.* with regard to household expenses. The manner in which a household, or a person's private expenditure, is ordered. A society ordered after the manner of a family.

b. In a wider sense: The administration of the concerns and resources of any community or establishment with a view to orderly conduct and productiveness; the art or science of such administration.

c. *Political Economy.* Originally the art or practical science of managing the resources of a nation so as to increase its material prosperity; in more recent use, the theoretical science dealing with the laws that regulate the production and distribution of wealth.

ENEMY: a. an unfriendly or hostile person.
b. One that cherishes hatred, that wishes or seeks to do ill to another.

ENVIRONMENT: a. The objects or the region surrounding anything.
b. The conditions under which any person or thing lives or is developed; the sum total of influences which modify and determine the development of life or character.

EPIPHANY: a. Manifestation, striking appearance, esp. an appearance of a divinity.
b. A manifestation or appearance of some divine or superhuman being.

ETHICS: a. The science of morals; the department of study concerned with the principles of human duty.
b. The moral principles by which a person is guided.
c. In wider sense: The whole field of moral science, including besides Ethics properly so-called, the science of law whether civil, political, or international.

EVIL adj.: a. The antithesis of GOOD in all its principal senses *In OE this word is the most comprehensive adjectival expression of disapproval, dislike or disparagement. In quite familiar speech the adj. Is commonly superseded by **bad**: the sb. is somewhat more frequent but chiefly in the widest senses, the more specific senses being expressed by other words, as **harm, injury, misfortune, disease**, etc.*

Sb. b. That which is evil. In the widest sense: that which is the reverse of good.

c. What is morally evil; sin, wickedness.

FACT:

a. Something that has really occurred or is actually the case; something certainly known to be of this character; hence, a particular truth known by actual observation of authentic testimony, as opposed to what is merely inferred, or to a conjecture or fiction; a datum of experience, as distinguished from the conclusions that may be based upon it.

b. That which is of the nature of a fact; what has actually happened or is the case; truth attested by direct observation or authentic testimony; reality.

FIELD:

a. Open land as opposed to woodland; as stretch of open land.

b. An area of sphere of action, operation, or investigation; a (wider or narrower) range of opportunities, or of objects, for labor, study, or contemplation; a department or subject of activity or speculation.

c. The space or range within which objects are visible through an optical instrument in any one position.

d. *Physics.* The area or space under the influence of, or within the range of, some agent. *Magnetic field*: any space possessing magnetic properties, either on account of magnets in its vicinity, or on account of currents of electricity passing through or round it.

HARMONY:

a. Combination or adaptation of parts, elements, or related things, so as to form a consistent and orderly whole; agreement, accord, congruity.

b. Agreement of feeling or sentiment; peaceableness, concord.

c. Combination of parts or details in accord with each other, so as to produce an aesthetically pleasing effect; agreeable aspect arising from apt arrangement of parts.

d. The combination of musical notes, either simultaneous or successive, so as to produce a pleasing effect; melody; music, tuneful sound.

e. Pleasing combination or arrangement of sounds, as in poetry or in speaking; sweet or melodious sound.

HUBRISTIC: Insolent, contemptuous.

JUST: a. That does what is morally right, righteous, *Just before (with) God* or, simply, *just*: Righteous in the sight of God; justified. Now chiefly as a Biblical archaism.

b. Upright and impartial in one's dealings; rendering every one his due; equitable.

c. Consonant with the principles of moral right or of equity; righteous; equitable; fair. Of rewards, punishments, etc.: Deserved, merited.

d. Constituted by law or by equity, grounded on right, lawful, rightful; that is such legally.

e. Having reasonable or adequate grounds; well-founded.

f. Conformable to the standard, or to what is fitting or requisite; right in amount, proportion, aesthetic quality, etc.; proper; correct.

INDIGENOUS: Born or produced naturally in a land or region; native or belonging naturally *to* (the soil, region, etc.) (Used primarily of aboriginal inhabitants or natural products.)

JUSTICE: I. The quality of being just.

a. The quality of being (morally) just or righteous; the principle of just dealing; the exhibition of this quality or principle in action; just conduct; integrity, rectitude. (One of the four cardinal virtues.)

b. *Theol*. Observance of the divine law; righteousness; the state of being righteous or 'just before God'.

c. Conformity (of an action or thing) to moral right, or to reason, truth, or fact; rightfulness; fairness; correctness; propriety.

II. Judicial administration of law or equity.

a. Exercise of authority or power in maintenance of right; vindication of right by assignment of reward or punishment; requital of desert.

b. The administration of law, or the forms and processes attending it; judicial proceedings.

KARMA:

in Buddhism, the sum of a person's actions in one of their successive states of existence, regarded as determining their fate in the next; hence, necessary fate or destiny, following as effect from cause.

KNOWLEDGE:

a. The fact of knowing a thing, state, etc. or (in general sense) a person; acquaintance; familiarity gained by experience.

b. Acquaintance with a fact; perception, or certain information of, a fact or matter; state of being aware or informed; consciousness (of anything). The object is usually a proposition expressed or implied: e.g. the knowledge that a person is poor, knowledge of his poverty.

c. Intellectual acquaintance with, or perception of, fact or truth; clear and certain mental apprehension; the fact, state, or condition of understanding.

d. Acquaintance with a branch of learning, a language, or the like; theoretical or practical understanding of an art, science, industry, etc.

LANGUAGE:

a. The whole body of words and of methods of combination of words used by a nation, people or race; A 'tongue'.

b. In generalized sense: words and the methods of combining them for the expression of thought.

MILITARISM:

The spirit and tendencies characteristic of the professional soldier; the prevalence of military sentiment or ideals among a people; the political condition characterized by the predominance of the military class in government or administration; the tendency to regard military efficiency as the paramount interest of the state.

MONEY:	a. Current coin; metal stamped in pieces of portable form as a medium of exchange and measure of value. b. Coin considered in reference to its value or purchasing power; hence, property or possessions of any kind viewed as convertible into money or having value expressible in terms of money.
MORAL:	a. Of or pertaining to character or disposition, considered as good or bad, virtuous or vicious: of or pertaining to the distinction between right and wrong, or good and evil, in relation to the actions, volitions, or character of responsible beings; ethical. b. *Moral sense*: The power of apprehending the difference between right and wrong, *exp.* when viewed as an innate and unanalysable faculty of the human mind. c. *Moral virtue:* occasionally occurs in contradistinction to the 'Christian virtues" (Faith, Hope, Charity), or as restricted to such virtues as may be attained without the aid of religion.
PARADIGM:	A pattern, exemplar, example.
PEACE:	a. Freedom from, or cessation of, war or hostilities; that condition of a nation or community in which it is not at war with another. b. A state or relation of peace, concord, and amity. c. Freedom from civil commotion and disorder; public order and security. d. Freedom from disturbance or perturbation (esp. as a condition in which an individual person is); quiet tranquility, undisturbed state. e. Freedom from quarrels or dissension between individuals; a state of friendliness; concord, amity. f. Absence of noise, movement, or activity; stillness, quiet; inertness.
PHILOSOPHER:	a. A lover of wisdom; those who devote themselves to the search of fundamental truth; one versed in philosophy or engaged in its study; formerly in

a wide sense, including those learned in physical science (Physicists, scientists, naturalists), as well as those versed in the metaphysical and moral science, but now chiefly confined to the latter. Also with a defining word, as *moral philosopher, political philosopher, natural philosopher* (= physicist).

b. *The Philosopher*, spec. applied to Aristotle.

PHILOSOPHY: a. (in the original and widest sense). The love, study, or pursuit of wisdom, or of knowledge of things and their causes, whether theoretical or practical.

b. (= *metaphysical philosophy*). That department of knowledge or study which deals with ultimate reality, or with the most general causes and principles of things. (Now the most usual sense.)

c. Sometimes used especially of knowledge obtained by natural reason, in contrast with revealed knowledge.

d. A particular system of ideas relating to the general scheme of the universe; a philosophical system or theory.

e. The system which a person forms for the conduct of life. The mental attitude or habit of a philosopher; serenity under disturbing influences or circumstances; resignation; calmness of temper.

PUNISH: a. As an act of a superior or of public authority: To cause (an offender) to suffer for an offence; to subject to judicial chastisement as retribution or requital, or as a caution against further transgression; to inflict a penalty on.

b. To requite or visit (an offence, etc.) with a penalty inflicted on the offender; to inflict a penalty for (something).

c. To inflict punishment.

d. To handle severely; to inflict heavy damage, injury, or loss on.

PUNISHMENT: a. The action of punishing or the fact of being punished; the infliction of a penalty in retribution for an offence; also, that which is inflicted as a penalty; a penalty imposed to ensure the application and enforcement of a law.

RELIGION: a. Action or conduct indicating a belief in, reverence for, and desire to please, a divine ruling power; the exercise or practice of rites or observances implying this.
b. A particular system of faith and worship
c. Recognition on the part of man of some higher unseen power as having control of his destiny, and as being entitled to obedience, reverence, and worship; the general mental and moral attitude resulting from this belief, with reference to its effect upon the individual or the community; personal or general acceptance of this feeling as a standard of spiritual and practical life.

SCIENCE: a. A branch of study which is concerned either with a connected body of demonstrated truths or with observed facts systematically classified and more or less colligaged by being brought under general laws, and which includes trustworthy methods for the discovery of new truth within its own domain.

SPIRIT: The animating or vital principle in humans (and animals); that which gives life to the physical organism, in contrast to its purely material elements; the breath of life.

SPIRITUALITY: The quality or condition of being spiritual; attachment to or regard for things of the spirit as opposed to material or worldly interests.

SPIRITUAL: Of or pertaining to, affecting or concerning, the spirit or higher moral qualities, esp. as regarded in a religious aspect.

TERROR: a. The state of being terrified or greatly frightened; intense fear, fright, or dread.

b. The action or quality of causing dread; terrific quality, terribleness; also *concr.* A thing, or person that excites terror; something terrifying.

TERRORISM:	**a.** Government by intimidation as directed and carried out by the party in power in France during the Revolution of 1789-94,
b. A policy intended to strike with terror those against whom it is adopted; the employment of methods of intimidation; the fact of terrorizing or condition of being terrorized.

TERRORIST:	**a.** Any one who attempts to further his views by a system of coercive intimidation; *spec.* applied to members of one of the extreme revolutionary societies in Russia.
b. Dyslogistically: One who entertains, professes, or tries to awaken or spread a feeling of terror or alarm; an alarmist, a scaremonger.

TERRORIZE:	**a.** To fill or inspire with terror, reduce to a state of terror; esp. to coerce or deter by terror.
b. To rule, or maintain power, by terrorism; to practise intimidation

TRAGEDY:	A play or other literary work of a serious or sorrowful character, with a fatal or disastrous conclusion: opp. to COMEDY.* See Other Definitions

TRAGIC:	a Of, pertaining, or proper to tragedy as a branch of the drama; of the nature of tragedy; composing, or acting in, tragedy: opp. to COMIC.
b. Resembling tragedy in respect of its matter; relating to or expressing fatal or dreadful events; connected with or excited by such events; sorrowful, sad, melancholy, gloomy.
c. Resembling the action or conclusion of a tragedy; characterized by or involving 'tragedy' in real life; calamitous, disastrous, terrible, fatal.

TRUTH:	a. Conformity with fact; agreement with reality; accuracy, correctness, verity (of statement or thought).
	b. Agreement with a standard or rule; accuracy, correctness.
	c. True statement or account; that which is in accordance with the fact;
	d The fact or facts; the actual state of the case; the matter or circumstance as it really is.
	e. Conduct in accordance with the divine standard; spirituality of life and behavior.
	f. That which is true, real, or actual (in a general or abstract sense); reality; *spec.* in religious use, spiritual reality as the subject of revelation or object of faith.
TRUTHFUL:	Of persons (or their attributes): Disposed to tell, or habitually telling, the truth; free from deceitfulness; veracious.
TRUTHFULNESS.	a. The quality of being truthful.
	b Disposition to tell the truth; veracity.
	c. Accuracy in representing the reality; freedom from pretence or counterfeit, as in a work of art or literature.
WAR:	a. Hostile contention by means of armed forces, carried on between nations, states, or rulers, or between parties in the same nation or state; the employment of armed forces against a foreign power, or against an opposing party in the state.
	b. *transf.* and *fig.* Applied *poet.* or rhetorically to any kind of active hostility or contention between living beings, or of conflict between opposing forces or principles.
UNDERSTAND:	a. To comprehend; to apprehend the meaning or import of; to grasp the idea of.
	b. To be thoroughly acquainted or familiar with (an art, profession, etc.); to be able to practise or deal with properly.

c. To comprehend by knowing the meaning of the words employed; to be acquainted with (a language) to this extent.

d. To take or accept as a fact, without positive knowledge or certainty; to get as an impression or idea; to believe. Chiefly with obj. clause.

UNDERSTANDING: a. (Without article.) Power or ability to understand; intellect, intelligence.

b. *Of understanding*, intelligent, capable of judging with knowledge.

c. Of persons (or animals): Possessed of understanding; having knowledge and judgement; intelligent.

UNIVERSE: The whole of created or existing things regarded collectively; all things (including the earth, the heavens, and all the phenomena of space) considered as constituting a systematic whole, exp. as created or existing by Divine power; the whole world or creation; the cosmos.

WISDOM: a Capacity of judging rightly in matters relating to life and conduct: soundness of judgement in the choice of means and ends; sometimes, less strictly, sound sense, *esp.* in practical affairs: opp. to *folly.*

b. Personified (almost always as feminine).

c. *Pl.* As attribute of a number of persons; hence, with possessive, as a title of dignity or respect, esp. for the members of a deliberative assembly.

d. Knowledge (*esp.* of a high or abstruse kind); enlightenment, learning, erudition; in early use often = philosophy, science.

Other Definitions:

AHIMSA: nonviolence, harmlessness

HEARTSONG: Your inner beauty, the song in your heart that wants you to help make yourself a better person, and to help other people do the same. Everybody has one (Mattie Stephanek)

HUBRIS:	a. In the actions of mortals, is a deviation from the mean of good and right conduct, or acting in harmony with one's conscience or the Life Force; the greatest deviation being the killing of another human being. b. The attempt to violate or overrule the cosmic force, the primal unifier of the universe.
SPECIEL:	Pertaining to the species.
TRAGEDY:	a. Originating in Greek drama, **Tragedy,** *known as classical tragedy*, in its highest sense, is the one where ' calamitous, disastrous, terrible, fatal' human events are averted at the very moment that the loss is felt on the pulses. b. An imitation of human beings in action, representing a microcosm of the cosmos. c. **Tragedy** can reveal the remorseless workings of the universe in a way that humans can learn from it and adjust their behavior to avert, even at the last moment, cataclysmic as well as personal loses and dehumanization.

(From Aristotle's Poetics)

INDEX

This index has been prepared to the best of our ability. It is intended to encourage readers and users of this book to reflect upon the wide range of contributors to the ideals of our contemporary civilization. We also ask for your indulgence where it is not complete.

Note: The exact location of many quotations is not known. However where the particular passage can be identified, that poem or article has been included, in italics. Where a book has been particularly studied for use in any of the articles, it has also been noted, in italics.

Adams, Henry Brooks. 1838 - 1918. Historian, teacher, editor

Adams, Jane. 1860 - 1935. Founded Hull House, whose projects include the first juvenile court in U.S. Supported woman suffrage, peace activist & Nobel Peace Prize winner in 1931.

Addison, Joseph. 1672 - 1719. English essayist, poet and statesman.

Adler, Felix. 1851 - 1933. Educator and ethical reformer. *An Ethical Philosophy of Life.*

Adler, Peter S. Explorer of meditation alternatives to environmental litigation; represents the Western Justice Center.

Aeschylus. 525 - 456 B.C. E. Eldest of the three great Greek tragedians.

Aikman, William. 1682 - 1731. Scottish portrait painter.

Aleyn, Charles. D. 1640. English historical poet.

Allen, Fred (real name John Florence Sullivan) 1894 - 1956. American comedian.

Amiel, Henri Frederic, 1821 - 1881. Swiss poet and philosopher.

Andrews, Lynn. Devoted her life to study of age-old rituals and shamanism.

Apocrypha. Biblical books in the Greek version of the Old Testament, not in the Hebrew Bible. Word coined by St. Jerome in fifth century.

Aquinas, Saint Thomas. 1225? - 1274. Italian scholastic philosopher.

Arden, Harvey. Senior writer for National Geographic magazine. Editor with Steve Wall, *Wisdomkeepers. Meetings with Native American Spiritual Elders.*

Aristotle. 384 - 322 B.C. Greek philosopher.

Armstrong, Karen. English born prolific author, teacher, commentator and foremost scholar on Religious Affairs. *Islam*

Arnold, Matthew. 1822 - 1888. English poet and critic.

Athenagoras. Athenian Christian apologist of the 2nd Century.

Augustine, Saint. (Latin name Aurelius Augustinus). 354 - 430. Philosopher; bishop of Hippo.

Aurelius, Marcus. In full **Marcus Aurelius Antoninus.** See **Marcus Aurelius**

Aurobindo, Sri. 1872 - 1950. Born in Calcutta, author; founder, Sri **Aurobindo Ashram;** visionary for Auroville (First Universal City in India).

Bab, 1819 - 1950. Forerunner of the Baha'i Faith.

Bach, Richard. Truth seeker. *Jonathan Livingston Seagull.*

Baha'u'Lla'h (Baha'i). Real name Mirza (i.e. Prince) Husayn Ali. 1817 - 1892. Persian religious leader, born in Teheran. Founder of Baha'i Faith.

Bah'a, Abdu'l. 1844 - 1921. Interpreter, Baha'i Faith.

Bailey, Alice. 1880 - 1949. Author, founder of Lucis Trust and World Goodwill.

Baker, Dorothy Gillam. Author. *Climax of History from Turbulence to Transformation.*

Balzac, Honore de. 1799 - 1850. French novelist.

Bancroft, George, 1800 - 1891. American historian.

Barrymore, Ethel, 1879 - 1959. American actress.

Barth, Joseph. 1906 - American Unitarian clergyman.

Bear Paw. Native American leader..

Beaumont, Frances, 1584 - 1616. English playwright.

Beauvoir, De, Simone 1900 - 19 French author *The Second Sex*

Beecham, Sir Thomas. 1879 - English conductor and impressario.

Behn, Aphra. 1640 - 1689. English dramatist and novelist. First English woman professional writer.

Beranger, Pierre Jean de. 1780 - 1857. French poet.

Bergman , Ingmar. 1918 - Prolific Swedish theatre director, playwright, screenwriter, oscar winner.

Bernanos, Georges. 1888 - 1948. French novelist, essayist, soldier in W.W.I., later known as the 'bard of the French Resistance'. *Plea for Liberty.*

Beversluis, Joel. Ed. *Sourcebook of the World's Religions.*

*Bhagavad Gita***:** Major Hindu scripture, somewhat parallel to Christian New Testament.

Bhandara. Mobed Zarir. High Priest, Zoroastrian Association of California.

Bible: Acts 18:10: Amos 5: 24; Corinthians 11. Ecclesiastes 3:12; 11:13. Ephesians 4:32; 32. Hebrews 11:1. Isaiah 30:15. Jeremiah 29:13; 33:3.

John 1:9; 4:18. Luke 1:9; 12:27; 12:31; 22:27. Matthew 6:21; 6:22; 20; 6:34; 7:6; 11:28; 16:25. Philippians 1:6; 4:8.Proverbs 3:6; 23:7. Psalms 11:10; 60:12. Romans 8:28.

Black Elk. 1863 - 1950. . Native American medicine man, spiritual leader. *Earth Prayer*.

Blake, William, 1757 - 1827. English artist, poet, mystic. *Tiger*.

Block, Rudolph (real name Bruno Lessing). 1879 - 1940. American author.

Bodhidhama. Died about 530 C.E. Buddhist monk from southern India; went to China (520) as Buddhist missionary known as first Buddhist patriarch of China; founded contemplative school of Buddhism in China

Bogan, Louise. 1897 - 1970. American poet. *After the Persian*

Bohr, Niels 1885 - 1962. Danish physicist. Adapted the quantum theory to atomic structure. 1922 Nobel prize in physics.

Bolen, Jean Shinoda. Author of *Goddesses in Everywoman, Gods in Everyman*; psychiatrist, Jungian analyst.

Bolitho, William (full name: William Bolitho Ryall). 1891 - 1930. British journalist and author.

Borge, Victor. 1909 - 2000. Humorist, entertainer.

Bovee, Christian Nestell. 1820 - 1904. American author.

Bowring, Sir John. 1792 - 1872. English consular agent and linquist, author of poems and hymns.

Bradbury, Ray. American novelist, short story writer, playwright, screenwriter. Science fiction author.

Bradstreet,Anne. 1612-1672. English poet who migrated to Massachusetts, became author of first collection of original poems produced in America. *In Reference to her Children*.

Brahma Kumaris: Worldwide spiritual movement with headquarters at Mount Abu, India.

Brandeis, Louis Dembitz, 1856 - 1941. American jurist.

Bridges, Robert. 1844- 1930. English poet. *All Beauteous Things*.

Bright, John. 1811 - 1889. Englist orator and statesman.

Bromfield, Louis, 1896 - 1956. American author.

Bronte, Emily, 1818 - 1848. English novelist and poet. *No Coward Soul*.

Brooke, Medicine Eagle. Intertribal Indian metis, an Earth-keeper, singer, ceremonial leader, healer, teacher dedicated to learning from Mother Earth/Father Spirit.

Browning, Elizabeth Barrett, 1806 - 1861. English poet. *The Cry of the Children, Is God not with us?*

Browning, Robert. 1812 - 1889. English poet. *Pippa's Song*.

Bryson, Breeze. Songwriter. *One Great Moment, Light the Torch.*

Buck, Pearl. 1892 - 1973. American novelist devoted to mutual understanding between the peoples of Asia and the U.S.

Buddhism: One of the world's major religions; began in India and spread to the rest of Asia and beyond.

Bui, Hum, M.D., Developer of a CaoDai Center in Pomona, California.

Burbank, Luther. 1849 - 1926. Horticulturalish.

Burke, Edmund, 1729 - 1779. British statesman and orator.

Bushnell, Horace. 1802 - 1876. American congregational clergyman.

Cadillac, Sieur Antoine de la Mothe. 1658 - 1730. French administrator in America, governor of Louisiana 1713-16.

Cameron, Anne. Internationally respected author of fiction, screen-plays and poetry. *Daughters of Copper Woman.*

Carey, Ken. Inspirational author. *The Third Millennium.*

Casals, Pablo. 1876 - 1973. Spanish violoncellist, conductor, composer, professor.

Cavendish, Margaret, Duchess of Newcastle, 1623 - 1674. Poet. *The Hunting of the Hare.*

Chagall, Marc, 1887 - 1985. Russian painter.

Channing, William Ellery. American clergyman, Unitarian/Universalist.

Chardin, Pierre Teilhard de. See **Teilhard de Chardin**

Charron, Pierre. 1541 - 1603. French Roman Catholic theologian and philosopher.

Chatwin, Bruce. 1942 - 1989. British author chronicler of Australian aborigines' struggle to preserve sacred lands. *The Songlines.*

Chopra, Deepak. Indian popular writer, lecturer, New age guru, Chopra Center for Well Being.

Christianity: One of the world's major religions; began in Palestine and spread worldwide.

Churchill, Sir Winston Leonard Spencer 1874 - 1965. British statesman, author, prime minister.

Cicero, Marcus Tullius. 106 - 43 B.C. Roman orator, statesman and philosopher.

Coleridge, Samuel Taylor. 1772 - 1834. English poet and critic, interpreter of the meaning of the imagination in the movement of the organic universe.

Colton, Charles Caleb. 1780? - 1832. English Anglican clergyman and sportsman.

Confucianism: A major religion/philosophy which is centered in China.

Confucius, 551? - 479? B.C.E. Chinese philosopher.

Congreve, William. 1670 - 1729. English playwright.
Conow, Hoff..Author. *A Resurgent Model of the Universe*.
Copernicus, Nicolaus. 1473 - 1543. Polish astronomer. Regarded as founder of modern astronomy, for the theory that earth rotates on its axis and planets revolve around the sun.
Cousins, Norman. 1915 - 1990. Research writer, editor of *Saturday Review*, author of books on Schweitzer and Gandhi.
Crane, Frank. 1861 - 1928. American clergyman and journalist.
Curie, Marie. (Marja Sklodowska). 1867 - 1934. Physical chemist

Daijin. Related to Konko church, a derivation from Shinto.
Dante, Alighieri. 1265 - 1321. Italian poet.
Das, Tulsi. 1532 - 1623. A Sarwariya Brahman most famous Hindi poet.
Delacampagne, Christian. Graduate of the Ecole Normale Superieure in Paris. Teacher at Connecticut College, New London. Ed. *A History of Philosophy in the Twentieth Century*.
De Quincey, Thomas. 1785 - 1859. English author.
Dewey, John. 1859 - 1952. American philosopher and educator.
Dickens, Charles John Huffam. 1812 - 1870. English novelist. Toured America advocating international copyright and abolition of slavery.
Dickinson, Emily Elizabeth. 1830 - 1886. American poet.
Diderot, Denis. 1713 - 1784. French encyclopedist, philosopher.
Dolby, Cheryl. Stained glass artist, teacher, shop-owner, sculptor. *She Who Whispers*.
Donne, John. 1573 - 1631. English poet.
Dior, Christian. 1905 - 1957. World famous fashion designer.
Dryden, John, 1631 - 1700. English poet, dramatist.
Duffy, Al. Buddhist tour guide in Southern California area.
Durant, Will. (William James). 1885 - 19.. American editor, author.
Dyer, Wayne. Motivational speaker, author. Formerly with Stanford Research Institute.

Effendi, Shoghi. 1897 - 1957. Defender of the Faith. Baha'i.
Einstein, Albert. 1879 - 1955. American physicist, born in Germany. Swiss citizen, U.S. Citizen.
Eisenhower, Dwight David. 1890 - 1969. American army general, 34th president of the U.S.
Eisler, Riane. Author of *The Chalice and the Blade, Our History Our Future*. Co-President of Center for Partnership Studies, Malibu.

Elgin, Duane. Speaker, author. Formerly with Stanford Research Institute.

Eliot, George - Mary Ann Evans. 1819 - 1880. English novelist.

Elizabeth I. 1533 - 1603. A Renaissance woman, encouraged great writers, had Shakespeare's plays performed in England's first theatre. Published an Elizabethan Prayer Book.

Ellwood, Robert S. and Partin, Harry B. *Religious and Spiritual Groups in Modern America.*

Emerson, Ralph Waldo. 1803 - 1882. American essayist and poet.

Epictetus. Approx. 90 C.E. Greek Stoic philosopher.

la Farge, John. 1911 - 1963. Jesuit priest urged better racial relations in the U.S. Joined the 1963 March on Washington.

Fell, Margaret. 1614 - 1702. Born Margaret Askew, m. Thomas Fell in 1632. Her home was the early organizational headquarters of the Friends.

Fiske, Stephen Longfellow. American award-winning singer/songwriter, dedicated to peace, environmental and humanitarian concerns. *The Art of Peace.*

Fitzgerald, Edward. 1809 - 1883. English poet and translator. *Rubaiyat of Oma Khayyam.*

Flewelling, Ralph Tyler. 1871 - 1960. American philosopher.

Follett, Mary Parker. 1868 - 1933. A quaker born in Massachusetts.

Forrest, Nathalie LaCroix, D.D., Autoharpist, minister, singer. *Save Our World.*

Foster, Michelina. Guitarist, songwriter metaphysician. *Unity-'n-Diversity.*

Frank, Anne. 1929 - 1944. Author. Immortalized the human face of the holocaust. *The Diary of a Young Girl.*

Frankfurter, Felix. 1882 - 1965. American jurist, born in Austria. U.S. Supreme Court 1939-1965.

Franklin, Benjamin. 1706 - 1790. American statesman, scientist and author.

Frost, Robert. 1874 - 1963. American poet, professor.

Frothingham, Octavius Brooks. 1822 - 1895. Unitarian and independent minister.

Fuller, Margaret. 1810 - 1850. American writer, spent time in Italy. *Woman in the Nineteenth Century.*

Funch, Flemming. Founder of New Civilization Network.

Galileo, Galilei. 1564 - 1642. Italian astronomer and physicist. Believed the earth and the planets revolve around the sun.

Gandhi, Mahatma. (Mohandas Karamchand). 1869 - 1946. Hindu nationalist leader, champion of non-violence.

Garfield, James Abram. 1831 - 1881. Twentieth president of the United States.

Gawain, Shakti. Leader in world consciousness movement; author, Co-Founder of New World Library.

George, David Lloyd, See **Lloyd George, David**.

Gibran, Kahlil. 1883 - 1931. Syrian poet, painter and author. *The Prophet*.

Gide, Andre. 1869 - 1951. French author and critic.

Gilligan, Carol. Professor, Graduate School of Education, Harvard University. *In a Different Voice*.

Gladstone, William Ewart. 1809 - 1898. British statesman.

Goethe, Johann Wolfgang von. 1749 - 1832. German poet.

Goldman, Emma. 1869 - 1940. Russian anarchist, lived in America.

Goldsmith, Joel S. Religious science writer.

Gournay, Marie Le Jars. 1565 - 1645. Author. *Equality of Men and Women*.

Gray, Asa. 1810 - 1888. American botanist.

Greene, Brian. Rhodes scholar, professor of Physics and Mathematics, visionary in the quest for a theory of everything. *The Elegant Universe*.

Greene, Theodore Meyer. *The Arts and the Art of Criticism*.

Green, Glenda. Nationally acclaimed Australian artist, former University Professor and spiritual author.

Grenfell, Sir Wilfred Thomason. 1849 - 1906. English Baptist minister and explorer in Africa.

Guillen, Michael. Holds a Ph.D. in Physics, Mathematics and Astronomy. Science editor at ABC News.

Guiney, Louise Imogen. 1861 - 1920. American poet, lived in England.

Gurumayi. Spiritual leader. A Siddha Guru, she also writes books and songs for children.

Hammarskjold, Dag. 1905 - 1961. Swedish statesman; Secretary General of United Nations 1953-1961.

Hanh, Thich Nhat. 1926 - Buddhist inspirational guru and author from Vietnam.

Hawken, Paul. Author connected with Findhorn community in Scotland.

Hazlitt, William. 1778 - 1830. English essayist and critic.

Herson, Rabbi Benjamin. Rabbi Emeritus, Malibu Jewish Center and Synagogue; Founder Wallenberg Institute of Ethics.

Hills, Burton. Spiritual author.

Hinduism: One of the world's major religions, centered in India.

Holmes, Ernest. 1887 - 1960. Founder of Religious Science; minister.

Hope, Alec Derwent. 1907 - 19.. Australian poet.

Hopkins, Emma Curtis. 1853 - 1925. Student of Mary Baker Eddy. Founded E.C.H. College of Metaphysical Science. Encouraged women to take leadership roles. *Scientific Christian Mental Practice.*

Hopkins, Gerard Manly. 1844 - 1889. English poet , professor, Jesuit priest. *Pied Beauty*

Hornaday, Dr. William. Senior minister of Founders Church of Religious Science.

Hosmer, Frederick Lucine. Songwriter. *Hear, Hear, O You Nations.*

Hroswitha of Gandersheim. 935 - 1001. First recorded European dramatist after the Dark Ages.

Hubbard, Barbara Marx. World Futurist emphasizing fulfilment, not destruction; author, speaker.

Hubbard, L. Ron. 1911 - 1986. American philosopher and humanitarian. Founder of Scientology.

Hughes, Langston. 1902 - 1967. Artist. Black community leader.

Hume, David. 1711 - 1776. Scottish philosopher, historian.

Huntington, Frederick Dan. 1879 - 1930. First Protestant Episcopal Bishop of Central New York. Author.

Hurston, Zora Neale. 1891 - 1960. One of the writers of the Harlem Literary Renaissance. *How it Feels to Be Colored Me.*

Huxley, Aldous Leonard. 1894 - 1963. English author.

Huxley, Thomas Henry. 1825 - 1895. English biologist.

Ibsen, Henrik. 1828 - 1906. Norwegian poet and playwright.

Ickes, Harold Le Clair. 1894 - 1952. American lawyer and politician.

Irving, Washington. 1783 - 1859. American author.

Islam: One of the world's major religions. Originated in Saudi Arabia.

Jackson, Jill Songwriter. *Let There be Peace on Earth.*

Jackson, Michael. Songwriter. *We are the World.*

Jainism: A religion of India, emphasized ahisma (harmlessness).

Jalal-ud-Din, Rumi. 1207 - 1273. Persian poet, oriental mystic; founded order of dervishes, a Sufi sect.

James, Henry. 1843 - 1916. American novelist.

James, Patricia. Mertaphysician, songwriter. The New Age is Now.

James, William. 1842 - 1910. American psychologist, philosopher.

Jayanti, Sister. Global Cooperation House, London, Brahma Kumaris World Spiritual University.

Jeans, Sir James Hopwood. 1877 - 1946. English physicist and author.

Jefferson, Thomas. 1743 - 1826. Third president of the United States.

Jerrold, Douglas William. 1803 - 1857. English playwright, editor, humorist.

Jesus. Founder of Christian religion; lived in Palestine.

Joachim of Fiore. 1135 - 1202. Italian Cisterian monk. *Age of the Spirit*.

Joad, Cyril Edwin Michinon. 1891 - 1953, English philosopher.

Johnson, Samuel. 1709 - 1784. English author, lexicographer, conversationalist.

Jordan, June. 1936 - 2002. Professor African American Studies and Women's Studies at the University of California, Berkeley. *In Memoriam: Poem for Mrs. Fannie Lou Hamer*.

Jose, Arthur Wilberforce. 1863 - 1934. English born Australian historian.

Joubert, Joseph. 1754 - 1824. French moralist.

Judaism: One of the world's major religions. Began in Palestine before Christianity.

Julian of Norwich. 1342 - 1423?. English mystic. *Showings (Revelations)*.

Juline, Kathy. Editor, *Science of Mind Magazine*.

Junius, Franciscus. 1589 - 1677. German-born philologist. Published editions of *Caedmon* and preserved Anglo-Saxon manuscripts.

Kant, Immanuel. 1724 - 1804. German philosopher.

Karenga, Maulana. 1941 - Cultural activist and scholar. Created *Kwanzaa,* the African American and Pan-African holiday

Keats, John. 1795 - 1821. British romantic poet. *Ode to a Grecian Urn*.

Keen, Sam. Pioneer in personal mythology, twenty year editor of *Psychology Today*, international lecturer.

Keller, Wolfgang, Rev. Dr. Church of Scientology of Los Angeles.

Kennedy, John Fitzgerald. 1917 - 1963. Thirty fifth president of the United States.

Kennedy, Robert Francis. 1925 - 1968. American lawyer, United States attorney general, senator.

Kermode, Josephine "Cushag". English poet. *John the Priest*

Khan, Hazrat Inayat. Sufi leader.

Kilmer, Alfred Joyce. 1886 - 1918. American poet killed in action, World War 1.

Kimball, Richard. Professor at University of the West in Buddhist Studies.

King, Marthin Luther Jr. 1929 - 1968. American civil rights leader.

Kipling, Rudyard. 1865 - 1936. Author, poet, winner of 1907 Nobel Prize, Literature. *If.*

Knox, Richard. Songwriter. *Amen.*

Kornfield, Jack. Author, *Buddha's Little Instruction Book.*

Krishna, Lord. A well known Hindu god.

Krishnamurti, J. 1895 - 1986. A theosophically trained but independent spiritual teacher.

Kubler-Ross, Elizabeth. Death and dying medical doctor. Has center for children dying of AIDS.

Kucinich, Dennis J. 1954 - Ohio congressman, opposed war in Iraq 2003, ran for President 2004, advocating the establishment of a *Department of Peace.*

Kuznetsky, Richard, M.D. Regional Representative, Bawa Muhaiyadeen Fellowship.

Lama, The Dalai. 1935 - Tibetan spiritual leader and former political leader of Tibet; Nobel Peace Prize 1989.

Lamb, Charles & Mary. 1764 - 1847. English poets. *A Child.*

Lamennais, Felicite Robert de. 1782 - 1854. French priest and philosopher.

Lanier, Sidney. 1842 - 1881. American poet.

Laplace, Marquis Pierre Simon de. 1749 - 1827. French astronomer and mathematician.

Larson, Christian D. Influential early New Thought leader. In 1908 organized the New Thought Temple, author of 'The Optimist Creed'.

Laski, Harold Joseph. 1893 - 1950. English political scientist and educator.

Laszlo, Erwin. Professor, Science Adviser to the Director General of UNESCO., Editor in Chief *The World Encylopaedia of Peace. The Limits of Mankind.*

Lee, Barbara. Bay Area Congresswoman, voted against the war in Iraq in 2003. Her main concerns are housing, education and the environment.

Lehman, Herbert Henry. 1878 - 1963. American banker and politician.

Leighton, Shirley. Minister, author, Founder of the First Faith of the Future.

Leo XIII. (Gioacchino Vincenzo Pecci). 1810 - 1903. Pope 1878-1903.

Lincoln, Abraham. 1809 - 1865. Sixteenth president of the United States.

Link, Henry Charles. Religious author.

Lisieux, St. Theresa of. 1873 - 97. French Carmelite nun, canonized in 1925. Known as the 'Little Flower of Jesus' and 'Doctor of the Church'.

Lloyd George, David. 1863 - 1945. British prime minister 1916 - 22.

Longfellow, Henry Wadsworth. 1807 - 1882. American poet.

Longinus, Dionysius Cassius. .? - 273. Greek philosopher *On the Sublime, On Great Writing*

Lowell, Amy. 1874 - 1925. American poet.

MacArthur, Douglas. 1880 - 1964. American general.

MacDonald, George. 1824 - 1905. Scottish novelist.

Mails, Thomas E. Lutheran pastor and renowned author of many books on Native American individuals and cultures. *The Hopi Survival Kit.*

Mandela, Nelson. First president of the new South Africa, internationally renowned moral and spiritual leader, winner of 1993 Nobel Peace Prize.

Mann, Mary Anneeta, Ph.D. Playwright, author. *The Construction of Tragedy. (Preface)*

Mann, Stella Terrill. Spiritual author.

Marcus Aurelius, surnamed **Antoninus.** Original name Marcus Annius Verus. 161 - 180. Roman emperor, Stoic philosopher.

Marmontel, Jean Francois. 1723 - 1799. French writer, author of tragedies.

Marquis, Donald Robert Perry (Don). 1878 - 1937. American journalist and humorist.

Marsden, John. Australian poet.

Marshall, Peter. Presbyterian minister and evangelist, spiritual writer.

Maslow, Abraham. Spiritual psychologist, author, foremost spokesman of the humanistic movement, philosopher of science, teacher of self-actualization.

Massinger, Philip. 1583 - 1640. English playwright.

Matheson, George. Songwriter. *Gather Us In, O Love.*

Matthews, Clifford N., Mary Evelyn Tucker, Philip Hefner, ed. *When Worlds Converge.*

Maurois, Andre, *pseudonym of* Emile Salomon Wilhelm Herzog. 1885 - 1918. French author, liaison officer with British forces during World War I.

Maxwell, Elsa. 1883 - 1963. Born in Iowa, lived in California. Famous as a party-giver.

McTaggart, Lynne.. Award winning journalist, visionary in understanding the coalescence of paradigms in all fields of human endeavor. *The Field.*

Mead, Margaret. 1901 - 1979. Renowned American anthropologist and writer, examined and explained other cultures.

Mencken, Henry Louis. 1880 - 1956. American editor and satirist.

Mernissi, Fatema. Foremost scholar and thinker in the Middle East, teaches sociology at University Mohammed V in Rabat, Morocco. *Islam and Democracy.*

Meynell, Alice, (nee **Thompson**). 1850 - 1922. English poet and essayist. *The Shepherdess.*

Michelangelo. (Michelangelo Buonarroti). 1475 - 1564. Italian sculptor, painter, architect, poet.

Millay, Edna St. Vincent. 1892 - 1950. American poet, first woman to receive Pulitzer Prize for Poetry. *God's World.*

DeMille, Agnes. 1905 - 1993. Choreographer, dancer, screenwriter, producer.

Miller, Sy. Songwriter. *Let There be Peace on Earth.*

Milton, John. 1608 - 1674. English poet.

Mindess, Harvey. 1902 - American Behavioral Scientist, wrote about humor and the meaning of life.

Mitchell, Maria. 1819 - 1889 Professor of Astronomy, Vassar College. First discoverer of a telescopic comet. Led the 1875 Women's Congress.

Montesquieu, Baron de La Brede. 1689 - 1775. French lawyer, author and philosopher.

Moore, Marianne. 1887 - 1972. American poet winner of Pulitzer Prize.

Morgan, Marlo. American health-care professional, and traveler to Australia's outback. *Mutant Message Down Under.*

Morley, Christopher Darlington. 1890 - 1957. American author.

Mormon. A member of the Church of Jesus Christ of Latter Day Saints, a movement which began in the United States in the nineteenth century.

Muir, John. 1838 - 1914. American naturalist, born in Scotland.

Muller, Robert. 1923 - Born in Alsace-Lorraine, joined the United Nations in 1948 becoming an Assistant Secretary. Visionary for Humanity's First Millennium of Peace. *New Genesis: Shaping a Global Spirituality.*

Mumford, Lewis. 1895 - 1990. American author, architect, philosopher.

Murray, W. H. 1913 - 1986. Founding trustee of the John Muir Trust, Scottish mountaineer, profuse writer concerning the need for protecting and conserving landscapes.

Nader, Ralph. Consumer advocate, lawyer, leader of Green Party, outspoken leader for political reform.
Native American: The indigenous peoples of the United States.
Nehru, Jawaharlal 1889 - 1964. Indian nationalist: prime minister 1947-1964.
Neruda, Pablo. 1904 - 1973. Chilian poet, Nobel Prize for Literature 1971. Born Neftali Ricardo Reyes Basoalto.
Newman, John Henry. 1801 - 1890. English theologian, cardinal.
Nietzsche, Friedrich Wilhelm. 1844 - 1900. German philosopher and poet.

Ogden, C.K. and Richards, I.A. *The Meaning of Meaning.*
O'Keefe, Georgia. 1887 - 1986. Poet, inspirational writer.
Oppenheimer, J. Robert. 1904 - 1967. Appointed Director of the Manhattan Project in 1942, he oversaw the creation of the atomic bombs dropped on Hiroshima and Nagasaki in 1945.
Ovid (Publius Ofidius Naso). 43 B.C.E. - ?17 C.E. Roman poet.

Pascal, Blaise. 1623 - 1662. Friench scientist and philosopher.
Pasternak, Boris Leonidovich. 1890 - 1960. Russian poet, novelist.
Patterson, Andrew Barton (Banjo). 1864 - 1941.Australian poet. *Clancy of the Overflow*
Paul, Pope VI. Catholic leader.
Pauling, Linus. 1901 - 1994. American chemist and scientist, winner of Nobel prizes for Chemistry and Peace, father of the first Nuclear Test Ban Treaty. *No More War.*
Peale, Norman Vincent. 1898 - 1993 American clergyman, author, known as the apostle of self-esteem.
Perry, Grace. 1927 - 1987. Australian poet and editor of *Poetry Magazine. Black Swans at Berrima.*
Perry, Ralph Barton. 1876 - 1957. American philosopher and educator.
Pitter, Ruth. 1897 - 1992. English poet, recipient of the Queen's Gold Medal for Poetry 1955.
Pius XII. Catholic pope.
Plato. 427? - 347 B.C.E. Greek philosopher.
Po Chu-I. 772 - 846. Chinese poet, most prolific of the Tang poets.

Pollock, Channing,   1880 - 1946.   American novelist, playwright, lecturer.
Pope, Alexander. 1688 - 1744. English poet.
Proust, Marcel. 1871 - 1922. French novelist.

Quran. The major Islamic scripture.

Raine, Kathleen. 1908 - 2003 English poet of Scottish ancestry, one of the poets who founded the *Temenos Review* in 1980. *The Unloved*.
Raleigh, Sir Walter. 1552? - 1618. English courtier, navigator, historian and poet.
Ram, N. Sri. - 1963. Theosophical spiritual teacher.
Ramakrishna, Sri. Founder of Vedanta Society.
Reeves, Marjorie. Distinguished educationalist and scholar. *Joachim of Fiore and the Prophetic Future*.
Renee, Louise. San Francisco City Attorney.
Richie, Lionel. Songwriter. *We Are the World*.
Rinpoche, Sogyal. Renowned teacher, author. *The Tibetan Book of Living and Dying*.
Roberts, Jane. 1929 - 1984. Poet, mystic, known for Seth Material.
Robertson, Frederick William.   1816 - 1853.   English Anglican clergyman.
Robinson, Roland. 1912 - 1992. Irish born Australian 'outback' poet, transcriber of Australian Aboriginal myths. *Aboriginal Myths and Legends*.
Romaine, William. 1714 - 1795. English Calvinist theologian.
Roosevelt, Eleanor. 1884 - 19.. Lecturer, writer, sponsor of UNESCO. *U.N. General Assembly 12.12.52*.
Roosevelt, Franklin Delano. 1882 - 1945. Thirty second president of the United States.
Rossetti, Christina Georgina. 1830 - 1894. English poet. *Up-Hill*.
Rousseau, Jean-Jacques. 1712 - 1778. French philosopher, author, born in Switzerland.
Rowe, Nicholas. 1674 - 1718. English poet and playwright.
Rubenstein, Arthur. 1887 - 1982. Polish born internationally renowned pianist.
Rukeyser, Muriel. 1913 - 1980. American poet. *More of a Corpse Than a Woman*.
Rumi, Jelaluddin, see Jalaluddin Rumi.
Ruskin, John. 1819 - 1900. English art critic and author.

St. Vincent Millay, Edna. 1892 - 1950. American poet.

Salk, Jonas Edward. 1914 - 1995. Scientist, founder of Salk Institute, developed first vaccine against poliomyelitis.

Sappho (of Lesbos). About 600 BC. Greek lyric poet.

Saraydarian, Torkom. Born in Asia Minor. Violinist, teacher, lecturer, mechanical engineer, meteorologist, writer and philosopher. *Hiawatha and the Great Peace*.

Schaef, Anne Wilson. International teacher, facilitator, writer, philosopher.

Schwarz, John. String theory pioneer at California Institute of Technology.

Schweitzer, Albert. 1875 - 1965. French clergyman, philosopher, physician, musician.

Secker, Thomas. 1693 - 1768. English Anglican clergyman. Archbishop of Canterbury 1758-1768.

Secondat, Charles de. See Montesquieu, Baron de La Brede

Selye, Hans. 1907 - 1982. Canadian physician, author, born in Austria, considered the 'father' of stress hypothesis.

Seneca, Lucius Annacus. 4 B.C.E? - 65 C.E. Roman statesman, philosopher.

Shakespeare, William. 1564 - 1616. English playwright, poet.

Shankar, Sri Sri Ravi. Spiritual teacher.

Shaw, George Bernard. 1856 - 1950. British playwright, author, critic.

Shelley, Mary Wollstoncraft 1797 - 1851. English author.

Shelley, Percy Byssche. 1792 - 1822. English poet. *Ode to The West Wind*.

Shinto: Japanese religion.

Shirley, James. 1596 - 1666. English playwright.

Sikhism: A religion of India. Most Sikhs wear turbans.

Simmons, Charles. 1798 - 1856. American Congregational clergyman, author.

Sinetar, Marsha. Editor, author of *Ordinary People as Monks & Mystics*.

Sinnott, Edmund W. 1888 - 1968. Author, philosopher, writing about science in society.

Smiles, Samuel. 1812 - 1904. Scottish biographical author.

Socrates. 470? - 399 B.C. Greek philosopher.

Solon. 638? - ?559 B.C. Athenian lawgiver, one of the Seven Wise Men of Greece.

Sophocles 496? -406 BC. Greek tragic playwright.

Spangler, David. Author, Associate of Brugh Joy. *Song of the Avatars.*

Spinoza, Baruch. 1632 - 1677. Dutch philosopher.

Spring-Rice, Sir Cecil. 1859 - 1918. British diplomat, minister to Persia, ambassador to U.S.

Stepanek, Matthew (Mattie) J.T. 1990 - 2004. PEACEMAKER, poet. *Heartsongs,* et. al*

Stevenson, Adlai Eqing. 1900 - 1965. American lawyer and politician.

Stewart, Leland, B.S.E., S.T.B. Founder/Central Coordinator of the Unity-and-Diversity World Council. *World Scriptures.*

Stowe, Harriet Beecher. 1811 - 1896. Prolific writer, anti-slavery speaker in America and Europe.

Strehlow, Theodore. 1908 - 1978. Australian born linguist and ethnologist. Transcribed Australian Aboriginal legends and songs. *Songs of Central Australia.*

Sun, Patricia. Teacher of Creative Communication on worldwide television, radio at universities and churches.

Swaback, Vernon D,. FAIA, FAICP. An Australian visionary. *The Creative Community.*

Swetchine, Anne Sophie. 1782 - 1857. Russian-French author.

Tagore, Sir Rabindranath. 1861 - 1941. Poet, born in Bengal, Knighted in 1915.

Taoism: An ancient mystical religion of China.

Taylor, Jeremy. 1613 - 1667. English Anglican clergyman, author, Chaplain-in-ordinary to Charles I.

Teilhard de Chardin, Pierre. 1881 - 1955. French paleontologist and biologist, Jesuit priest.

Temenos Review. Founded in 1980.

Tennyson, (Alfred Lord). 1809 - 1892. English poet, poet laureate.

Teresa, Mother. 1910 - 1997. Sisters of Charity sister. Nobel Peace Prize, 1979. Considered for sainthood.

Thales. 640? - 546 B.C.E. Greek philosopher, scientist.

Thayer, William Makepeace. 1820 - 1898. American author..

Thoreau, Henry David. 1817 - 1862. American author and naturalist.

Tibetan Doctrine: Teachings from Tibet in *World Scriptures*.

Tolle, Eckhart. Modern author and lecturer on meditation.

Tolstoy, Leo. 1828 - 1910. Russian novelist, philosopher and religious mystic.

Trench, Richard Chenevix. Songwriter. *Make Channels for the Streams of Love.*

Truman, Harry S. 1884 - 19.. Thirty- third president of the United States.

Tutu, Archbishop Desmond, Nobel prize recipient 1984, Chairman of the Truth and Reconciliation Commission, Worldwide lecturer. *God Has a Dream.*

Tsalagi. A Native American language.

Tze, Lao. c604 - c521 B.C. Major teacher in Taoism.

Tzu, Sun. B.C.E. 2000? Chinese warrior - philosopher. *The Art of War,* a study of the anatomy of organizations in conflict.

Ueshiba, Morihei. 1883-1969. Japanese founder of Aikido. *The Art of Peace.*

Ullathorne, William Bernard. 1806 - 1889. English archbishop of the Roman Catholic Church.

UNICEF. United Nations Specialized Agency serving children.

Vandenberg, Arthur Hendrick 1884 - 1951. American senator, U.S. Delegate to the United Nations Conference in 1945 and Delegate to the General Assembly in 1946.

Van Gogh, Vincent. 1853 - 1890. Dutch Post-Impressionist painter. Artist.

Vauvenargues, Marquis de- Luc de Clapiers. 1715 - 1747. French soldier, Moralist. Wrote *Introduction a la Connaissance de l'Esprit Humain.*

Vivekananda, Swami. Author and worldwide spokesperson for Vedanta.

Voltaire, *assumed name of* **Francois Marie Arouet.** 1694 - 1778. French author.

Wall, Steve. Documentary photographer primarily for *National Geographic* magazine. Editor with Harvey Arden, *Wisdomkeepers. Meetings with Native American Spiritual Elders.*

Warren, Earl. 1891 - 19.. American jurist, chief justice, U.S. Supreme Court.

Washington, George. 1732 - 1799. First president of the United States.

Webb, Mary Gladys (nee Meredith). 1881 - 1927. English novelist, poet. *The Spirit of Earth.*

Webster, Daniel. 1782 - 1852. American lawyer, statesman, orator.

Wells, Herbert George. 1866 - 1946. English novelist, historian.

Wharton, Edith, (nee **Jones**). 1862-1937. American novelist.

Whitehead, Alfred North. 1861 - 1947. English mathematician, philosopher, forerunner in reviving philosophical describing of the organic universe. *Science and the Modern World.*

Whitman, Walt. 1819 - 1892. American poet.

Whittier, John Greenleaf. 1807 - 1892. American poet.

Wickham, Anna. 1884 - 1947 English poet with an Australian upbringing. *The Singer*

Wilcox, Ella Wheeler. 1850 - 1919. American journalist, poet. *Immortality.*

Wilcox, Dave. 1942 - Outstanding football player.

Willis, Nathaniel Parker. 1806 - 1867. American editor, author.

Wilson, Edward O. Natural scientist, father of bio-diversity culminating in his book on the unity of knowledge, *Consilience.*

Wilson, (Thomas) Woodrow. 1856 - 1924. Twenty-eighth president of the United States.

Wolf, Fred Alan, Ph.D., Consulting physicist, writer and lecturer exploring quantum physics, religion, philosophy and spirituality. *The Spiritual Universe.*

Wollstonecraft, Mary. 1759 - 1797. Writer of books urging the equality of men and women. *Vindication of the Rights of Woman.*

Wordsworth, William. 1770 - 1850. English poet.

Wright, Judith. 1915 - 2000. Australian poet, winner of the Queen's Medal for Poetry, 1992.

Wright, Frank Lloyd. 1869 - 1959. American architect.

Wylie, Elinor. 1885 - 1928. American poet.

Yiddish Proverb

Yogananda, Paramahansa. 1893 - 1952. Spiritual guru, founder of Self Realization Fellowship.

Yung, Carl Gustave. 1875 - 1961. Holistic psychologist, founder of Analytical psychology.

Ywahoo, Dhyani, Native American author, lecturer. *The Peoples of the Fire.*

Zoroastrianism: An ancient religion from Persia; predates Judaism.

*Heartsong: your inner beauty, the song in your heart that wants you to help make yourself a better person, and to help other people do the same. Everybody has one.

ABOUT THE AUTHORS/COMPILERS, AND NONPROFIT ORGANIZATIONAL SUPPORT

AUTHORS/COMPILERS

Mary Anneeta Mann, Ph.D.

Descended from Australian pioneers, Mary A. Mann has a Ph.D. in Communications and Drama from the University of Southern California. Her books include *HUBRIS, The Construction of Tragedy*, analyzing six tragedies from the Greeks to the present day in the light of the Science of Being; *Anzac*, a play about war and peace in the 20th century; two editions of *The Los Angeles Theatre Book*; *ThuGun and Natasha*, **a drama with rap,** *moving beyond guns and violence; Two Family Plays: Maria and the Comet* and *The Round Table*. She is the dramatist-in-residence at the Synthaxis Theatre Company and has been a member of Unity-and-Diversity for the past 14 years. As a member of the editorial board for *Science and Spirituality* and the *Spiritual Celebration Guide*, she researched contributions from many of the great scientists, artists and philosophers of our civilization, expanding the timeline, creating the index and preparing the manuscript for publication.

Rev. Leland Stewart, B.S.E., S.T.B.

Rev.Leland Stewart is the Founder/Central Coordinator of the Unity-and-Diversity World Council. He is also the author/compiler of *World Scriptures*, an interpretation of the world's religions and spiritual movements, designed for use in interfaith worship and meditation, as well as to facilitate mutual understanding around the world . Leland wrote *From Technological Power to Lasting Peace*, which is a philosophy of the emerging global civilization. He is a member of the editorial board of the *Science and Spirituality* book and its predecessor, the *Spiritual Celebration Guide*. One of his particular contributions to the *Science and Spirituality* book is the notes on the world's religions and spiritual movements, which are adapted from *World Scriptures*.

COMPILERS

Monique Baudaux Andersen

Artist and painter, Monique Baudaux Andersen, is a graduate of the Institute Normal de Dessin in Brussels. She has been interested in the concepts of UDC and has been involved with the organization in various capacities for many years. As a member of the editorial board she sponsored the publication of the *Spiritual Celebration Guide* and has brought a European perspective to the selection of the quotations for *Science and Spirituality*.

Rene Crawford

Rene had a spiritual awakening at age thirteen and attended Bible school in her teens, later branching into studying extensively many of the great faiths. She has found truth in all religions and has found Unity-and-Diversity to be an excellent opportunity to associate with like-minded people with different backgrounds. She was a contributor to the *Spiritual Celebration Guide* and is an ardent supporter of the activities of Unity-and-Diversity World Council. She is currently very active with the Agape International Spiritual Center.

Shirley Leighton

Shirley Leighton was ordained in the Unity-and-Diversity Ministry in 1976. Since that time she has traveled extensively, connecting with many groups who stress integrity and purpose in our lives. Her writing includes poetry as well as inspirational articles which have been published in many house organs including the *Journal for Religion and Psychical Research* and the *Science of Mind Magazine* while she was on the staff there. As a student of world religions, her major contribution to the field is a brief description of many of the world's great religions and their founders in *Avatars, Saints and Saviors* which also includes a timeline showing where each fits into history chronologically. Now living near the mountain ranges of Tucson, Arizona, she continues to work toward publishing her own volumes. As a member of the editorial board for *Science and Spirituality*, Shirley contributed material to the timeline as well as many of her own quotations and others.

Rev. Elizabeth Stewart

Elizabeth is a former elementary school teacher, artist, intensive journal writer, and spiritual pilgrim. She is a co-founder and current Board

Member of Unity-and-Diversity World Council (UDC), as well as a leader in UDC's World Interfaith Network (WIN). She has served for six years on the Peace Sunday Steering Committee since the event began under the Council's sponsorship. She believes that war is not the answer, but that love and understanding can lead us beyond war into a harmonious global civilization. As a member of the editorial board, her special contribution to *Science and Spirituality*, and its predecessor the *Spiritual Celebration Guide,* is her work on the candlelighting readings. She originated this ceremony as an essential part of the **Interfaith Celebration.**

NONPROFIT ORGANIZATIONAL SUPPORT:

Unity-and-Diversity World Council (UDC)

Unity-and-Diversity World Council is a nonprofit, tax-exempt California corporation. It was originally formed to develop the ideals and activities undertaken by six member organizations during International Cooperation Year 1965, voted into being by the General Assembly of the United Nations for non-governmental organizations. Its intergroup work continues to expand and to move toward becoming a "Peoples United Nations".

Once a month an Interfaith Celebration involving the world's religions and modern spiritual movements is held at various houses of worship in Southern California, using the *Unity-and-Diversity Spiritual Celebration Guide.* These meetings are developed by World Interfaith Network (WIN), a Unity-and-Diversity specialized affiliate.

Other intergroup events include annual Peace Sundays as well as workshops, seminars, conferences, training intensives and various networking endeavors and action projects for over thirty-five years.

The beginnings of geographic councils exist in several parts of the world, including Ghana, India, and the San Francisco Bay Area.

Publications include *World Scriptures*, published in 2003 in hardcover and paperback, a fifty-three year project of founder Leland P. Stewart in cooperation with numerous faith groups and people in the science of consciousness and related fields. *A Unity-and-Diversity Spiritual Celebration Guide* was published in 1994, coordinated by Dr. Mary A. Mann.

Copies of both of these publications may be obtained by writing to Unity-and-Diversity World Council, P.O. Box 661401, Los Angeles, Ca., U.S.A. 90066-9201.